Chromatography—The Ultimate Analytical Tool II

Chromatography—The Ultimate Analytical Tool II

Guest Editors

Victoria Samanidou
Natasa Kalogiouri

MDPI

Basel • Beijing • Wuhan • Barcelona • Belgrade • Novi Sad • Cluj • Manchester

Guest Editors
Victoria Samanidou
School of Chemistry
Aristotle University of
Thessaloniki
Thessaloniki
Greece

Natasa Kalogiouri
School of Chemistry
Aristotle University of
Thessaloniki
Thessaloniki
Greece

Editorial Office
MDPI AG
Grosspeteranlage 5
4052 Basel, Switzerland

This is a reprint of the Special Issue, published open access by the journal *Molecules* (ISSN 1420-3049), freely accessible at: https://www.mdpi.com/journal/molecules/special_issues/chromatography_analytical_tool_II.

For citation purposes, cite each article independently as indicated on the article page online and as indicated below:

Lastname, A.A.; Lastname, B.B. Article Title. *Journal Name* **Year**, *Volume Number*, Page Range.

ISBN 978-3-7258-6786-8 (Hbk)
ISBN 978-3-7258-6787-5 (PDF)
https://doi.org/10.3390/books978-3-7258-6787-5

Contents

About the Editors

Victoria Samanidou

Victoria Samanidou is Full Professor and Director of the Laboratory of Analytical Chemistry in the School of Chemistry of Aristotle University of Thessaloniki, Greece. Since 2022 she is also vice president of the School of Chemistry. Her research interests focus on the development of sample preparation methods using sorptive extraction prior to chromatographic analysis in accordance with green chemistry demands. She has co-authored 235 original research articles in peer reviewed journals and 75 reviews, 96 editorials/in view/opinions/commentaries and 65 chapters in scientific books or tutorials (h-index 53, Scopus Author ID 7003896015, 10219 citations).

In the last 6 years, she has been included in the list of World Top 2% Scientists, for single years and career as well, published in PLOS Biology based on citations from SCOPUS.

She has been also included in the Power List of Analytical Science several times. She is editorial board member of several scientific journals and guest editor in special issues.

Natasa Kalogiouri

Natasa Kalogiouri is an Assistant Professor of Analytical Chemistry in the Laboratory of Analytical Chemistry, in the School of Chemistry at the Aristotle University of Thessaloniki (AUTH). She has published 81 research papers, 4 chapter books, and more than 70 conference announcements. Her research interests focus on the development of analytical methods using liquid or gas chromatographic analysis, including multidimensional chromatographic systems coupled to high resolution mass spectrometric detectors. She is specialized in omics analysis and the identification of unknown metabolites and has developed in-house chemometric tools for data mining (data analysis, modeling, classification, and forecasting). Furthermore, she is a specialist in green sample preparation, focusing on the use of green microextraction techniques for the selective extraction of analytes (natural compounds, organic contaminants pharmaceuticals, etc.), from food, biological, and environmental matrices

molecules

MDPI

Article

An Optimized HPLC-DAD Methodology for the Determination of Anthocyanins in Grape Skins of Red Greek Winegrape Cultivars (*Vitis vinifera* L.)

Natasa P. Kalogiouri [1,*], Christina Karadimou [2], Mary S. Avgidou [2], Elissavet Petsa [2], Emmanouil-Nikolaos Papadakis [3], Serafeim Theocharis [2], Ioannis Mourtzinos [4], Urania Menkissoglu-Spiroudi [3] and Stefanos Koundouras [2,*]

1 Laboratory of Analytical Chemistry, Department of Chemistry, Aristotle University of Thessaloniki, 54124 Thessaloniki, Greece
2 Laboratory of Viticulture, School of Agriculture, Faculty of Agriculture Forestry and Natural Environment, Aristotle University of Thessaloniki, 54124 Thessaloniki, Greece
3 Pesticide Science Laboratory, School of Agriculture, Faculty of Agriculture Forestry and Natural Environment, Aristotle University of Thessaloniki, 54124 Thessaloniki, Greece
4 Laboratory of Food Chemistry and Biochemistry, Department of Food Science and Technology, Faculty of Agriculture, Aristotle University of Thessaloniki, 54124 Thessaloniki, Greece
* Correspondence: kalogiourin@chem.auth.gr (N.P.K.); skoundou@agro.auth.gr (S.K.)

Abstract: A rapid and simple HPLC-DAD analytical method was developed and optimized for the determination of anthocynanins in three red Greek winegrape varieties (Kotsifali, Limnio, and Vradiano). The critical parameters, such as the acidifying solvent and the extraction temperature, which affect the extraction of anthocyanins from the grapes, were studied to find the optimum values. The developed methodology was validated in terms of selectivity, linearity, accuracy, and precision and presented satisfactory results. The limits of quantification (LOQs) ranged between 0.20 mg/kg to 0.60 mg/kg, and the limits of detection (LODs) ranged between 0.06 mg/kg and 0.12 mg/kg. The RSD% of the within-day and between-day assays were lower than 6.2% and 8.5%, respectively, showing adequate precision. The accuracy ranged between 91.6 and 119% for within-day assay and between 89.9 and 123% for between-day assay. Sixteen samples from the main regions of each variety as well as from the official ampelographic collections of Greece were collected during the 2020 growing season and were further analyzed by HPLC-DAD. Notable differences in the anthocyanin content were detected among the cultivars using hierarchical cluster analysis (HCA).

Keywords: anthocyanins; HPLC; hierarchical cluster analysis; positive environmental footprint; Greek grape varieties; *Vitis vinifera*

1. Introduction

Phenolic compounds constitute a ubiquitous group of natural pigments from the flavonoid family widely distributed in fruits. These polyphenolic compounds are glycosides of polymethoxy and polyhydroxy derivatives of the 2-phenylbenzopyrylium or the flavylium ion [1]. The aglycone forms constitute the group of anthocyanidins. The most common anthocyanidins are delphinidin, cyanidin, malvidin, petunidin, peonidin, and pelargonidin. Grapes are rich in anthocyanidins with the exception of pelargonidin. Grape anthocyanidins belong to a diverse group of compounds called "secondary metabolites", which are synthesized principally as a plant adaptation to abiotic or biotic stresses, but which are also crucial for the quality of red wines, namely wine color intensity, hue, and stability.

Anthocyanins are located in the skins of grape berries, starting their accumulation at the veraison stage, a short lag phase separating two distinct periods of berry development: an initial phase characterized by rapid cell division and expansion in green berries and a second phase of growth by active solute accumulation, corresponding to grape ripening, at the end of which the anthocyanins reach their maximum levels. Thus, anthocyanins are important indicators in the determination of the harvest date.

Several agricultural factors exert a significant effect on the levels of grape anthocyanins. Natural factors such as the topography, soil, and climate of a vineyard location are reported to have a measurable impact on grape and wine color, mostly associated with their ability to induce different levels of vine growth and yield [2]. Seasonal operations (e.g., pruning, canopy manipulation, irrigation, fertilization, floor management) also affect the levels of anthocyanins by adapting the thermal and light conditions in the vine canopy [3]. However, genotype is the main factor differentiating the anthocyanin content of grapes and wines since the natural levels in berry skins are highly variable among *Vitis vinifera* cultivars. Furthermore, the profile of anthocyanins (relative abundance of individual anthocyanins, ratio of di-oxygenated vs. tri-oxygenated side-ring forms, ratio of acylated vs. non-acylated derivatives, etc.) is also variable among grapevine varieties [4,5]; therefore, it can be used as a chemotaxonomical criterion to establish differences between *Vitis vinifera* grape varieties [6] or other *Vitis* species [7].

The chemical characterization of grapevine cultivars is particularly important for the protection of minority varieties and the conservation of genetic diversity within the *Vitis vinifera* germplasm, which is endangered due to the generalized use of a small number of grape varieties in the international wine market [8]. Characterization and utilization of these varieties could also satisfy the increasing demand for new styles of wines by wine consumers [9]. Moreover, the profile of anthocyanins has technological and organoleptic repercussions on the winemaking process since it affects the intensity and stability of the red color in wine [10,11]. Color intensity increases with the number of substituted groups on the B-ring (di-oxygenated forms are redder while tri-oxygenated shift to blue) and with the replacement of hydroxyl by methoxyl groups (i.e., malvidin has the darkest color). Methoxylated anthocyanins (malvidin and peonidin) are also more stable than hydroxylated ones to environmental and viticultural factors [12]. Thereby, both the levels and the relative proportion of different anthocyanins in grape skins can confer distinctive characteristics to the produced wines depending on the cultivar; therefore, obtaining knowledge of the anthocyanic identity of each variety can provide a tool for applying the most appropriate agronomic and oenological techniques to maximize the varietal expression of the produced wines.

The determination of the anthocyanin content of grapes is a critical topic of the agricultural sector. Their bioactive content could be used as a chemical fingerprint in authenticity studies. Thus, there is a prominent need for the development of efficient, sensitive, and cost-effective analytical methodologies that could be applied in the determination of anthocyanins in grapes. In the literature, liquid chromatographic methods coupled to various detectors such as diode array (DAD) [13], fluorescence (FLD) [14], and mass spectrometric detectors (MS) [15] have been widely used in the analysis of anthocyanins [16]. Sample preparation is the first and most critical step of the analysis process. The applied extraction protocols have already been reviewed [13]. Solid–liquid extraction (SLE) is the most common technique used in the isolation of anthocyanins. According to the literature, extraction is commonly carried out using acetone or acidified methanolic solutions [13,17]. The optimal extraction conditions differ among the various plant materials since the efficiency of the extraction is affected by several parameters such as the type of the matrix, the chemical nature of the sample, the solvent used, the agitation method, the extraction time, the acidifying agent, and the temperature [17,18].

Greece is the cradle of a highly diverse grapevine genetic pool with more than 300 indigenous *V. vinifera* varieties, and most of them are confined in specific geographical areas. To the best of our knowledge, there have been few reports on the individual anthocyanin composition of Greek varieties [19,20]. In this study, we optimized the critical parameters of the extraction protocol for the investigation of the anthocyanin composition of three red-skinned indigenous varieties originating from different areas of Greece, namely 'Kotsifali', cultivated in the area of Heraklion in Central Crete island, 'Vradiano', cultivated in Evia (the second biggest island of Greece) in Central Greece, and 'Limnio', an ancient variety of North Aegean Sea cultivated in the island of Limnos and in the peninsulas of Chalkidiki in North Greece. To achieve this goal, a novel HPLC-DAD method was developed and validated for the determination of anthocyanins.

2. Results and Discussion

2.1. *Optimization of Extraction Parameters*

Different extraction systems have been proposed for the isolation of anthocyanins from the matrix [21–23]. Ultrasound-assisted extraction (UAE), microwave-assisted extraction (MAE), and supercritical fluid extraction (SFE), among others, have been reported for the extraction of anthocyanins from plant matrices. Among them, UAE has gained popularity owing to its high efficiency, rapidity, and low solvent consumption [24]. Even though UAE has been shown to increase the extraction yield of anthocyanins, it is difficult to establish a consistent extraction protocol for all the plant matrices since these analytes exist in various concentrations depending on the species/cultivar [25] and thus the extraction parameters have to be adjusted to the matrix under study. According to the literature, acetone has resulted in higher recoveries of anthocyanins [26]. The use of aqueous mixtures of acidified solvents has been shown to stabilize the anthocyanins [27]. In order to find the optimum parameters to increase the efficiency of the extraction, the most important factors such as the acidity of the solvent and the temperature during the extraction were optimized [17] using the One Variable at a Time (OVAT) approach [28]. For the optimization experiments, approximately 14 mg of freeze-dried grape skin was weighted in 2-mL dark Eppendorf tubes according to Pinasseau et al. [21]. During the extraction, the efficiency of the following mixtures was evaluated: (a) 0.05% trifluroacetic acid (TFA) in acetone:water (70:30, *v*/*v*); (b) 0.1% TFA in acetone:water (70:30, *v*/*v*); (c) 0.05% HCl in acetone:water (70:30, *v*/*v*); (d) 0.1% HCl in acetone:water (70:30, *v*/*v*). The extraction system with the highest recovery rate was selected as the optimum. The second parameter that was optimized was the extraction temperature after evaluating the efficiency of the extraction at 4 °C, 25 °C, and 60 °C.

2.1.1. Acidifying Solvent

The effects of acidifying the extraction solvent with 0.05% TFA, 0.1% TFA, 0.05% HCl, and 0.1% HCl were evaluated. According to the results presented in Table 1, the calculated recoveries ranged between 89.3 and 116.2%, showing that all the acidifying solvents tested could be successfully used for the extraction of anthocyanins from the grapes. The efficiency of using HCl as an acidifying solvent has already been shown [24,29,30]. The low pH of the extraction favors anthocyanins' extraction [26], and the findings of this work suggest that a pH over the range 1–2 is satisfactory for the extraction of anthocyanins both for HCl and TFA. To prevent degradation via hydrolysis, 0.05% TFA was chosen as the optimal acid for the extraction.

Table 1. Recovery (%) of anthocyanins extracted with different acidified solvents.

Acidifier	%R Dlp	%R Cyn	%R Pt	%R Pn	%R Mlv
0.05%TFA	91.0 ± 1.4	91.5 ± 1.0	101.5 ± 14.6	97.7 ± 4.4	99.0 ± 0.2
0.1%TFA	91.9 ± 2.0	91.4 ± 1.4	93.1 ± 1.2	99.5 ± 2.1	101.1 ± 3.3
0.05%HCl	110.8 ± 14.8	104.9 ± 12.2	111.3 ± 11.7	116.2 ± 16.9	114.8 ± 13.6
0.1%HCl	89.3 ± 5.8	87.5 ± 2.7	95.4 ± 3.5	96.3 ± 0.2	98.5 ± 2.8

Dlp: delphinidin-3-O-glucoside; Cyn: cyanidin-3-O-glucoside; Pt: petunidin-3-O-glucoside; Pn: peonidin-3-O-glucoside; Mlv: malvidin-3-O-glucoside

2.1.2. Extraction Temperature

Another critical parameter that has been demonstrated to affect the anthocyanin extraction yield is the temperature. Three extraction temperatures (4 °C, 30 °C, and 50 °C) were tested, and the extraction recoveries of the anthocyanins are presented in Table 2.

According to literature data, the elevated temperatures during the extraction increase the extraction yield, and 50 °C has been selected as the most appropriate temperature in many works; however, this was not the case in our work. Interestingly, the proposed extraction protocol is independent of the extraction temperature, as the extraction recoveries were acceptable, ranging between 88.5% and 104.9% in all cases [17]. The results showed that the temperature does not affect the extraction efficiency. To avoid decomposition [31], the lowest extraction temperature of 4 °C was selected as the optimum to conduct the experiments.

Table 2. Recovery (%) of anthocyanins extracted at different temperatures.

Temperature	Dlp	Cyn	Pt	Pn	Mlv
4 °C	97.9 ± 1.3	101.6 ± 1.0	102.6 ± 4.7	93.8 ± 4.4	99.8 ± 0.2
30 °C	90.0 ± 6.6	91.6 ± 5.2	88.5 ± 7.6	96.4 ± 2.2	91.7 ± 5.4
50 °C	98.6 ± 0.9	104.9 ± 3.0	99.2 ± 0.6	104.0 ± 2.2	99.1 ± 0.6

Dlp: delphinidin-3-O-glucoside; Cyn: cyanidin-3-O-glucoside; Pt: petunidin-3-O-glucoside; Pn: peonidin-3-O-glucoside; Mlv: malvidin-3-O-glucoside.

2.2. Method Validation Results

The optimized HPLC-DAD methodology was validated to assess the anthocyanin content of the grape skins, and all the analytical parameters, including the calibration curves, linear range, the coefficients of determination (r^2), accuracy and precision, limits of detection (LODs), and limits of quantification (LOQs), are presented (Tables 3–5). The calibration curves were all linear over the range LOQs—20 mg/kg with an r^2 above 0.99, proving that they can be used for the quantification of the anthocyanins. The LOQs were found to range between 0.20 mg/kg to 0.60 mg/kg, while the LODs were relatively low, with a range from 0.06 mg/kg to 0.20 mg/kg (Table 3). The RSD% of the within-day (n = 6) and between-day assays ($n = 3 \times 3$) were lower than 6.2% and 8.5%, respectively, showing adequate precision. The accuracy was assessed by means of the relative percentage of recovery (%R) at low, medium, and maximum concentration levels of 0.5, 5, and 20 mg/kg, and the results were acceptable over a range of 91.6–119% for the within-day assay (n = 6) (Table 4) and from 89.9 to 123% for the between-day assay ($n = 3 \times 3$) (Table 5).

Table 3. HPLC-DAD validation parameters.

Compound	Calibration Equation	Linear range (mg/kg)	r^2	LOD (mg/kg)	LOQ (mg/kg)
Dlp	y = 72,031.3x − 8267	LOQ-20	0.996	0.12	0.40
Cyn	y = 77,893.6 x − 6868	LOQ-20	0.999	0.10	0.30
Pt	y = 73,634.8x + 2305	LOQ-20	0.999	0.18	0.60
Pn	y = 68,782.6x + 1053	LOQ-20	0.999	0.06	0.20
Mlv	y = 55,442x + 1895	LOQ-20	0.999	0.10	0.30

Dlp: delphinidin-3-O-glucoside; Cyn: cyanidin-3-O-glucoside; Pt: petunidin-3-O-glucoside; Pn: peonidin-3-O-glucoside; Mlv: malvidin-3-O-glucoside.

Table 4. Repeatability results of the method estimated as recoveries (%R, n = 6) for the studied anthocyanins at three fortification levels.

Compound	Low Concentration (%R, $n = 3 \times 3$)	%RSD	Medium Concentration (%R, $n = 3 \times 3$)	%RSD	High Concentration (%R, $n = 3 \times 3$)	%RSD
Dlp	114	3.9	112	7.4	106	4.8
Cyn	105	7.5	96.7	7.9	91.6	7.3
Pt	113	6.5	111	6.5	106	6.2
Pn	119	6.2	118	5.8	107	7.1
Mlv	118	8.4	116	7.4	108	5.4

Dlp: delphinidin-3-O-glucoside; Cyn: cyanidin-3-O-glucoside; Pt: petunidin-3-O-glucoside; Pn: peonidin-3-O-glucoside; Mlv: malvidin-3-O-glucoside.

Table 5. Intermediate precision results of the method estimated as recoveries (%R, $n = 3 \times 3$) for the studied anthocyanins at three fortification levels.

Compound	Low Concentration (%R, $n = 3 \times 3$)	%RSD	Medium Concentration (%R, $n = 3 \times 3$)	%RSD	High Concentration (%R, $n = 3 \times 3$)	%RSD
Dlp	120	7.3	114	2.4	111	4.4
Cyn	114	5.0	94.7	2.3	89.9	3.8
Pt	118	4.7	107	3.1	110	6.1
Pn	103	9.6	119	4.1	123	1.4
Mlv	115	4.1	120	6.8	117	4.4

Dlp: delphinidin-3-O-glucoside; Cyn: cyanidin-3-O-glucoside; Pt: petunidin-3-O-glucoside; Pn: peonidin-3-O-glucoside; Mlv: malvidin-3-O-glucoside.

2.3. *Grape Sample Analysis*

2.3.1. Identification of Anthocyanins

Sixteen grape samples belonging to the varieties 'Kotsifali', 'Limnio', and 'Vradiano' were analyzed in triplicate ($n = 3$), and five anthocyanins were determined in all samples. The retention times (RTs) as well as the maximum wavelengths (λmax, nm) for each compound are reported in Table 6. Figure 1 presents a characteristic chromatogram of a real spiked sample at a 1 mg/kg concentration level. Supplementary Figure S1 details the comparison of the different non-spiked chromatograms of the studied grape varieties. All the anthocyanins appeared in all of the studied samples with notable differences in terms of the concentration but also the ratio between them as a characteristic of each variety.

Table 6. Chromatographic retention times and maximum wavelengths of anthocyanins.

Compound	RT	λ max (nm)
Dlp	14.3	516
Cyn	15.7	510
Pt	16.8	543
Pn	18.1	512
Mlv	18.8	520

Dlp: delphinidin-3-O-glucoside; Cyn: cyanidin-3-O-glucoside; Pt: petunidin-3-O-glucoside; Pn: peonidin-3-O-glucoside; Mlv: malvidin-3-O-glucoside

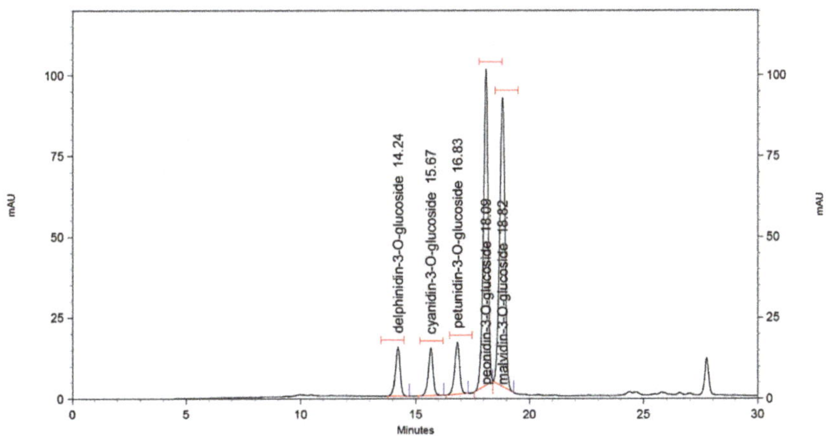

Figure 1. Characteristic chromatogram of a real sample spiked at 1 mg/kg concentration level, monitored at 520 nm.

2.3.2. Quantification Results

Significant differences in anthocyanin content were found between varieties, as each one has a specific set of anthocyanins that characterizes it [32]. This heterogeneity is mainly due to the effect of genotype [20]. In total, five anthocyanins were separated and quantified by the HPLC method in all samples, namely the glycosylated derivatives of delphinidin (Dlp), cyanidin (Cyn), petunidin (Pt), peonidin (Pn), and malvidin (Mlv). Apart from the five standard anthocyanins, another two peaks were detected. According to available literature [33–36], as the corresponding standards were not available, these two compounds should be malvidin-3-O-glucose acetate and p-coumarate, respectively.

Malvidin was by far the predominant anthocyanin in 'Limnio' grape samples, which is in agreement with several published studies about indigenous Greek varieties [4,7,37,38]. The quantitative determination of 'Kotsifali' and 'Vradiano' varieties showed that Mlv was equally important with Pn (Figure 2). Moreover, in the study of the Portuguese variety Alvarilhão [34], Mlv had a similar content to Pn. Two of the three varieties had comparable levels of Mlv ('Kotsifali' and 'Vradiano'), while the 'Limnio' variety appeared with up to two-times higher concentration (8.25–12.7 mg/100 g fresh weight (f.w.) in 'Kotsifali', 6.94–23.2 mg/100 g f.w. in 'Limnio', and 6.15–10.1 mg/100 g f.w. in 'Vradiano' grapes). According to the total anthocyanin quantification results that are graphically illustrated in Figure 3, the highest total anthocyanin concentration was observed in grape samples belonging to the 'Kotsifali' variety (49.2 mg/100 g f.w.), while the lowest was observed in samples belonging to the 'Limnio' variety (8.92 mg/100 g f.w.).

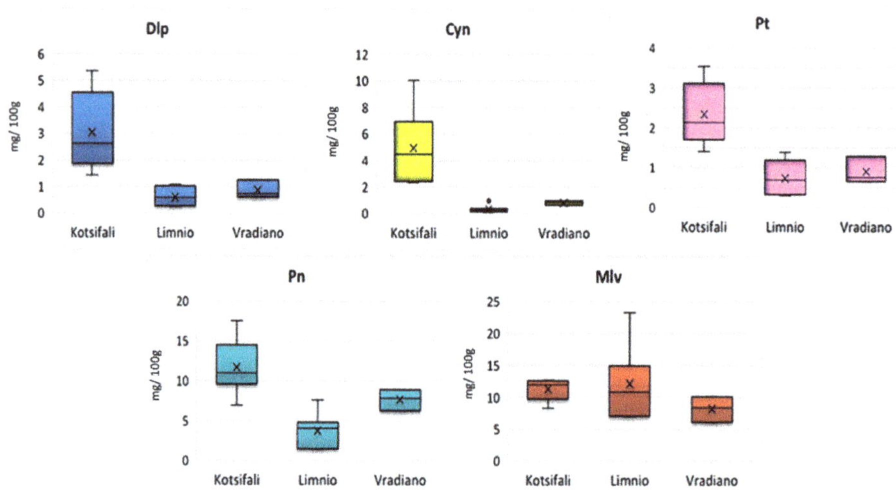

Figure 2. Box and whisker plots for the concentrations of Dlp; Cyn; Pt; Pn; Mlvin red grapes belonging to 'Kotsifali', 'Limnio', and 'Vradiano'.

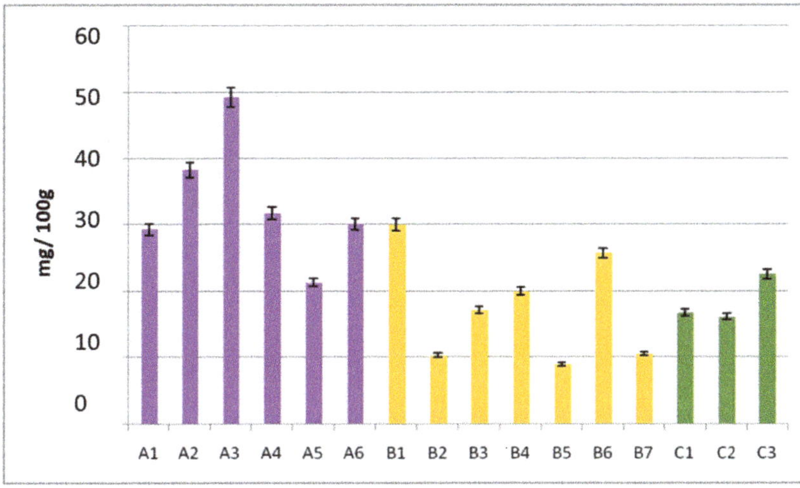

Figure 3. Total anthocyanin content (sum of individual anthocyanins ± SD) in red grapes of 'Kotsifali' (marked in purple), 'Limnio' (marked in yellow), and 'Vradiano' (marked in green).

Upon analyzing each monoglucoside separately, Cyn exhibited the lowest mean content (0.33 mg/100 g f.w. for 'Limnio' and 0.81 mg/100 g f.w. for 'Vradiano') followed by Dlp (0.60 mg/100 g f.w. in 'Limnio' and 0.87 mg/100 g f.w. in 'Vradiano'). The exception was the variety of 'Kotsifali', which presented a remarkably high concentration of Cyn when compared to the other varieties (4.93 mg/100 g f.w.). On the contrary, in 'Limnio' grapes, all anthocyanins except of Mlv had a small contribution to the total pool of anthocyanic content, while 'Vradiano' displayed an intermediate profile. All of the above are presented in detail in Table 7. These results are in agreement with previous findings for some of these varieties [36,39], except for 'Vradiano', a rare red grape, which has never been analyzed previously.

Table 7. Anthocyanin concentration levels in red grapes of 'Kotsifali', 'Limnio', and 'Vradiano' (samples analyzed in triplicate, $n = 3 \pm$ SD).

Variety	Sample Name	Dlp (mg/100 g)	Cyn (mg/100 g)	Pt (mg/100 g)	Pn (mg/100 g)	Mlv (mg/100 g)	Total (mg/100 g)
Kotsifali	A1	2.02 ± 0.04	2.43 ± 0.07	1.80 ± 0.04	10.4 ± 0.21	12.6 ± 0.23	29.3 ± 0.58
	A2	4.27 ± 0.16	5.9 ±0.07	2.96 ± 0.11	13.5 ± 0.32	11.6 ± 0.51	38.3 ± 1.17
	A3	5.35 ± 0.15	10.0 ± 0.11	3.54 ± 0.17	17.6 ± 0.50	12.7 ± 0.66	49.2 ± 1.59
	A4	2.88 ± 0.14	5.55 ± 0.33	2.16 ± 0.04	10.9 ± 0.13	10.2 ± 0.29	31.7 ± 0.92
	A5	1.42 ± 0.08	3.35 ± 0.13	1.41 ± 0.07	6.91 ± 0.26	8.25 ± 0.20	21.3 ± 0.74
	A6	2.38 ± 0.06	2.32 ± 0.11	2.12 ± 0.03	11.0 ± 0.17	12.3 ± 0.26	30.1 ± 0.72
Limnio	B1	1.07 ± 0.06	0.29 ± 0.01	1.39 ± 0.07	4.07 ± 0.20	23.2 ± 0.89	30.0 ± 1.23
	B2	0.26 ± 0.02	0.15 ± 0.01	0.33 ± 0.03	2.51 ± 0.17	7.08 ± 0.40	10.3 ± 0.64
	B3	0.59 ± 0.03	0.30 ± 0.01	0.68 ± 0.03	4.78 ± 0.11	10.8 ± 0.55	17.2 ± 0.71
	B4	0.74 ± 0.14	0.37 ± 0.05	0.84 ± 0.14	4.44 ± 0.57	13.6 ± 1.83	20.0 ± 2.70
	B5	0.23 ± 0.01	0.10 ± 0.01	0.29 ± 0.01	1.40 ± 0.12	6.94 ± 0.18	8.97 ± 0.32
	B6	1.03 ± 0.04	0.96 ± 0.16	1.19 ± 0.04	7.62 ± 0.62	14.9 ± 0.89	25.7 ± 1.75
	B7	0.31 ± 0.01	0.17 ± 0.01	0.37 ± 0.01	1.47 ± 0.06	8.22 ± 0.28	10.5 ± 0.36
Vradiano	C1	0.75 ± 0.02	0.64 ± 0.02	0.76 ± 0.02	6.25 ± 0.09	8.35 ± 0.17	16.7 ± 0.31
	C2	0.60 ± 0.03	0.84 ± 0.02	0.66 ± 0.03	7.88 ± 0.36	6.15 ± 0.25	16.1 ± 0.69
	C3	1.25 ± 0.16	0.96 ± 0.05	1.30 ± 0.16	8.90 ± 0.21	10.1 ± 0.64	22.5 ± 1.21

Dlp: delphinidin-3-O-glucoside; Cyn: cyanidin-3-O-glucoside; Pt: petunidin-3-O-glucoside; Pn: peonidin-3-O-glucoside; Mlv: malvidin-3-O-glucoside.

Relating the above results to percentages of each anthocyanin sum per variety (Figure 4), many notable differences emerged between the three varieties. Specifically, 'Limnio' skins' Mlv, which is the most stable of the five anthocyanins [40], was present at a range of 58–78.1% of the total. The respective percentage for 'Kotsifali' was 25.8–43%, while, for 'Vradiano', it was 38.1–49.9%. In both 'Kotsifali' and 'Vradiano', the contribution of Mlv was in the same level with Pn (32.4–36.5% in 'Kotsifali' and 37.3–48.8% in 'Vradiano'). The 'Kotsifali' variety also stood out due to its high percentage of Cyn (7.73–20.4%) when compared with the other studied varieties, wherein the latter anthocyanin was detected in traces. These results are in agreement with previous studies for this variety [4,20]. As mentioned previously, methoxylated anthocyanins, such as peonidin and malvidin, are more stable [20]. Consequently, the variety of 'Kotsifali', although rich in anthocyanins, is expected to have unstable color in the wines due to a high concentration of Cyn. The opposite is true for 'Limnio', with less but more stable color.

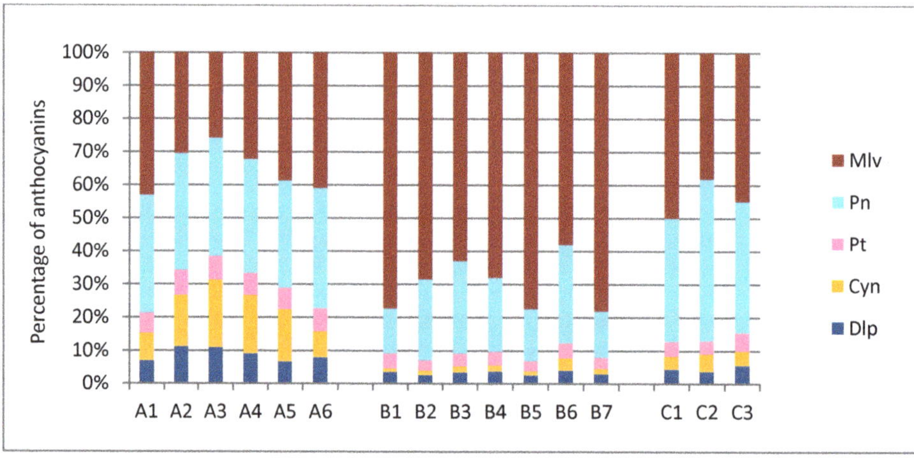

Figure 4. Percentage of individual anthocyanin content of Mlv; Pn; Cyn; Dlp; in the skins of the red grapes belonging to 'Kotsifali', 'Limnio', and 'Vradiano'.

These differences in both concentration and types of anthocyanin can lead to differences in intensity, hue, and color stability overall. Therefore, the knowledge of the anthocyanin identity of each variety can be a tool for the application of the most appropriate agronomic and oenological techniques to maximize the varietal expression of the produced wines.

2.4. Hierarchical Cluster Analysis

Hierarchical cluster analysis (HCA) was performed on the data matrix of 16 samples $\times$ 5 anthocyanins without a priori knowledge about the group structure of the dataset, measuring the distance between each pair of objects in terms of variables and grouping the objects that are close. HCA was applied to produce a tree diagram and identify the groups with objects of a high degree of similarity. The algorithm starts by treating each object as a singleton cluster (leaf); then, pairs of clusters are merged until all clusters have been successively merged into one large cluster that contains all the objects, resulting in a dendrogram.

The heatmap and the developed dendrogram of the HCA are presented in Figure 5, showing the clustering of three major groups, one for each variety.

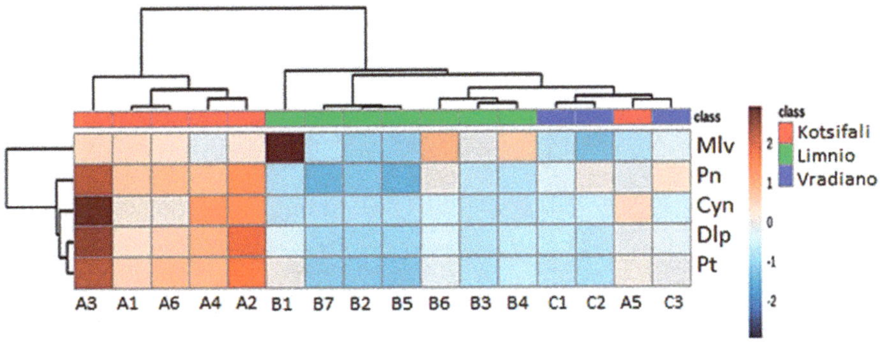

Figure 5. Hierarchical cluster analysis (HCA) of 16 samples of red grapes based on anthocyanins (Mlv; Pn; Cyn; Dlp; Pt).

Following the importance score, the A3 sample stands out due to its increased relativity to all of the monoglycoside anthocyanins. Similar behavior was displayed in all of the 'Kotsifali' samples except for the A5 sample, which was distinguished from the others by its high concentration at Cyn and low correlation with the others anthocyanins. The above results for the 'Kotsifali' variety, with the anthocyanins of Pn and Cyn standing out overall, are also confirmed in the heat map and the clustering of this variety. The anthocyanin composition of the variety 'Limnio' consists almost exclusively of Mlv, a fact confirmed by the dendrogram (Figure 5), with samples from the main cultivation areas of the variety (B6, B3, and B4) classified in the same sub-cluster. Lastly, Pn had an important percent in 'Vradiano' cultivar (37.3–48.8%) (Supplementary Table S1), which is characteristic of the variety and also becomes distinct in the heatmap (Figure 5). HCA analysis of the results showed (Figure 5) separation of the three varieties, particularly 'Kotsifali', from 'Limnio'. The 'Vradiano' variety differentiates, however, in the presence of the A5 sample because of Cyn concentration.

With the exception of A5, our results indicate that grape skin anthocyanin concentration ranges are characteristic of each variety and can therefore be used as a chemical indicator to distinguish Vitis vinifera varieties. Many studies using chemometric methods and especially HCA are able to classify wine and grapes according to grape variety, the phenolic content, or another variable. HCA analysis has been used to separate the cultivars and sort them based on skin color [41]. Clustering analysis has been employed for the discrimination of the Greek red wines belonging to the varieties 'Kotsifali' and 'Mandilaria' [42]. In another work, HCA was used for the clustering of red wines from China using the concentration levels of individual phenolic compounds [43].

3. Materials and Methods

3.1. Samples Collection

Grape samples belonging to the indigenous red grape of 'Kotsifali', 'Limnio', and 'Vradiano', originating from Greece, were collected at the stage of optimum maturity during the harvesting period of 2020. A representative sample was taken at the same time on the day for each cultivar from different vineyards throughout Greece depending on availability. Details about the variety and the geographical origin of the samples are presented in Table 8. Samples of 50 berries were collected for the grape maturity analysis. For the analysis of anthocyanins, another 50 berries were collected from each vineyard to form a bulk sample. Grape samples for anthocyanin analysis were put in boxes with dry ice and were brought to the laboratory and stored in a deep freezer (−80 °C) until further treatment.

Table 8. Variety and geographical origin of the samples.

Grape Variety	Sample No.	Geographical Origin	Location
Kotsifali	A1	Crete	Katw Asites
	A2	Crete	Dafnes
	A3	Crete	Alagni-Peza
	A4	Crete	Arxanes
	A5	Attica	Wine Institute
	A6	Macedonia	Greek Genebank
Limnio	B1	Attica	Wine Institute
	B2	Macedonia	Epanomi
	B3	Macedonia	Mount Athos
	B4	Macedonia	Sithonia
	B5	Thrace	Xanthi
	B6	Aegean Sea	Limnos
	B7	Macedonia	Serres
Vradiano	C1	Attica	Wine Institute
	C2	Evia	Istiaia
	C3	Evia	Gialtra

3.2. Sample Preparation

Most of 50 grapes were used for the calculation of total soluble solids (Brix), total titratable acidity (TA), and pH. These measurements are a part of grape maturity analysis (Supplementary Table S2). The samples for the anthocyanin analysis were weighted, and the skins were manually isolated. The skins were freeze-dried for two days and then were ground to obtain powder. The pulverized freeze-dried skins were stored in a deep freezer (−25 °C) until the analysis.

3.3. Reagents and Standards

Methanol and acetone (HPLC grade) were purchased from Merck (Zedelgem, Belgium). Methanol (HPLC-Ultra LC-MS) from HiPerSolv CHROMANORM, VWR Chemicals BDH (The Netherlands), was also purchased for the production of standard solutions. Formic acid (99%) and hydrochloric acid (37%) for analysis were purchased from Carlo Erba (Chaussée du Vexin, France), and trifluoroacetic acid for LC-MS was obtained from Fluka (Buchs, Switzerland). Standards of delphinidin-3-O-glucoside chloride, cyanidin-3-O-glucosidechloride, petunidin-3-O-glucoside chloride, peonidin-3-O-glucoside chloride, and malvidin-3-O-glucoside chloride were obtained from Extrasynthese (Genay Cedex, France). The standards of delphinidin-3-O-glucoside (Dlp), cyanidin-3-O-glucoside (Cyn), petunidin-3-O-glucoside (Pt), peonidin-3-O-glucoside (Pn), and malvidin-3-O-glucoside (Mlv) were diluted separately in methanol LC-MS with 0.1% HCl (concentration 1000 mg/L).

3.4. Instrumentation

Chromatographic analysis was carried out in a SpectraSYSTEM (Thermo Separation Products, Austin, TX, USA) HPLC system consisting of a P2000 secondary solvent pump, an AS3000 autosampler equipped with a 100-µL injection loop and a UV6000LP diode array detector. Chromatographic data were monitored and processed by ChromQuest 5.0 (Thermo Fisher Scientific Inc., Waltham, MA, USA). The samples were freeze dried in an Alpha 2–4 LD freeze dryer acquired from Martin Christ Gefriertrocknungsanlagen GmbH (Osterode am Harz, Germany) which was equipped with a two-stage vacuum rotary pump RZ 2.5 Vacuubrand (condenser temperature: −80 °C, max flow 2.3/2.8 m^3 h^{-1}, ultimate vacuum 4×10^{-4} mbar). A 5804 R centrifuge system with rotor F-45-30-11 was acquired from Eppendorf AG (Germany). Water was purified in a Direct-Q® 3 UV Water Purification System acquired from Merck KGaA, Darmstadt, Germany. For filtering the aqueous mobile phase, ME 25 ST 0.45-µm membrane filters (Schleicher and Schuell,

W. Germany) were used. For solvent evaporation under nitrogen gas, a TurboVap LV workstation was used by Caliper Life Sciences (Hopkinton, MA, USA). An ultrasonic bath RK 100H (Bandelin Sonorex, Berlin, Germany) and a Stuard-SB3 stirrer were used for the extraction. Regenerated Cellulose 0.22 µm (RC) syringe filters (Captiva, Agilent Technologies, Santa Clara, CA, USA) were used for filtering the samples.

3.5. Extraction of Anthocyanins and Anthocyanidins

Freeze-dried skin was weighted (0.0140 g) and extracted with 200 µL of methanol and 1.4 mL of acetone/water/TFA (70:29.95:0.05). The solution was extracted under sonication for 10 min and stirred at 40 rpm for 20 min, repeated twice. The extraction was carried out at 4 °C in the absence of light and the extract was centrifuged and thermostated at 4 °C, 10,000 rpm, for 10 min. Then, a fraction of 0.5 mL of the supernatant was dried with nitrogen under pressure. At this stage, and unless chromatographic analysis was followed immediately, the samples were stored as concentrates in the deep freezer (−80 °C). The concentrate was redissolved with a mixture of 250 µL methanol and 750 µL water 0.134% formic acid. Subsequently, the solution was led to an ultrasonic device again at 4 °C in the dark for 30 min. The extract was centrifuged and thermostated at 4 °C, for 14,000 rpm, for 15 min, and the supernatant was filtered through 0.22-µm RC syringe filters (RC) (Captiva, Agilent Technologies) prior to chromatography.

3.6. HPLC-DAD Analysis

The analysis of anthocyanins was performed using a modified chromatographic method according to Kyraleou et al. [22]. Chromatographic separation was performed on a Nucleosil 100-5 C_{18}, 250 × 4.6 mm, 5-µm, reversed-phase (RP) column (Macherey–Nagel, Düren, Germany). The DAD detector was set over the range 500–550 nm. The column oven temperature was 40 °C, the injection volume was 5 µL, and the total runtime was 40 min. The mobile phases were aqueous formic acid 5% (Solvent A) and methanol (Solvent B) at a flow rate of 1 mL/min. The gradient composition is presented in Table 9.

Table 9. HPLC Gradient for the separation of anthocyanins in grape skin extracts.

Time (Min)	Solvent A (%)	Solvent B (%)
0	90	10
22	50	50
32	5	95
34	5	95
35	90	10
40	90	10

3.7. Method Validation

The method validation was performed to estimate linearity, selectivity, LODs and LOQs, trueness, and precision. Linearity studies were performed in triplicate and covered the entire working range. The calibration curves of anthocyanins were constructed by plotting the peak area versus concentration. LODs were calculated as three signal to noise ratios (3 S/N), and the formula LOQ = 10 S/N was employed for the calculation of the LOQ [44]. Trueness and precision were studied using real grape skin samples spiked at three different concentrations (0.5, 5, and 20 mg/kg) and were analyzed in triplicate. To evaluate trueness, relative recoveries (%R) were calculated by means of recovery percentage by comparing the found and added concentrations of the examined analytes (mean concentration found/added concentration × 100). The precision of the method was expressed in terms of relative standard deviation (RSD%) and was calculated for repeated measurements of spiked samples. Following this approach, within-day precision (repeatability) was assessed in five replicates, while between-days precision (reproducibility) was assessed by performing triplicate analysis for spiked samples within four consecutive days [44].

In order to assess selectivity, five blank matrices were used, and no interferences were observed in the same chromatographic window as the anthocyanins examined.

3.8. Chemometric Analysis

The statistical differences between the species on the basis of their elemental concentration were estimated with ANOVA at a 95% confidence level ($p < 0.05$) in Microsoft Excel ((Microsoft, WA, USA) using the Data Analysis tool. The anthocyanin concentrations in grape samples were analyzed by hierarchical cluster analysis (HCA), which is a suitable method for small quantities of data, using the average between-groups linkage method and squared Euclidean distance interval measurement [45].

4. Conclusions

The purpose of this study was two-fold: first, to develop a simple and efficient methodology for the identification and quantification of grape skin anthocyanin in red cultivars, and, second, to investigate the anthocyanin profile of three Greek indigenous winegrape varieties as a means to distinguish cultivars based on chemometric analysis. For the preparation of the red grape skins to be analyzed, a rapid pretreatment protocol of the grape extracts was chosen, which combines simple and short purification techniques with a positive environmental footprint, due to the use of minimal amounts of solvents. The extraction protocol developed offers the possibility of the rapid identification and quantification of anthocyanins in order to characterize red varieties according to their anthocyanin profile. The grapevine variety 'Kotsifali' appeared richer in anthocyanins, albeit with a greater participation in the anthocyanin profile of the di-hydroxylated anthocyanins Pn and Cyn, while the opposite was observed in the 'Limnio' variety. This knowledge can improve the viticultural and oenological management of these varieties. Furthermore, the data revealed valuable information regarding the chemical separation of the wines of the three varieties based on chemometric analysis. These results create the conditions for further investigation in a large number of Greek grape varieties with the objective of establishing reference anthocyanin profiles to distinguish the varieties alone or in cooperation with molecular analyses.

Supplementary Materials: The following supporting information can be downloaded at: https://www.mdpi.com/article/10.3390/molecules27207107/s1, Table S1. Percentage of individual anthocyanin content in the skins of the red grapes varieties 'Kotsifali', 'Limnio', and 'Vradiano'; Table S2. Classical analysis results of red grapes; Figure S1. Characteristic chromatograms of real non-spiked samples, one of each variety: (a) Vradiano; (b) Kotsifali; (c) Limnio.

Author Contributions: Conceptualization, N.P.K., U.M.-S. and S.K.; methodology, C.K., E.P., M.S.A., S.T. and E.-N.P.; investigation, C.K., E.P., M.S.A., E.-N.P. and I.M.; data curation, N.P.K., U.M.-S. and S.K.; software, C.K. and N.P.K.; visualization, C.K. and N.P.K.; formal analysis, N.P.K., C.K. and S.K.; writing—original draft preparation, N.P.K. and C.K.; writing—review and editing, N.P.K., U.M.-S. and S.K.; supervision, N.P.K., U.M.-S. and S.K.; resources, U.M.-S. and S.K.; funding acquisition, U.M.-S. and S.K.; project administration, S.K.; All authors have read and agreed to the published version of the manuscript.

Funding: This research was funded by Greek national funds through the Public Investments Program (PIP) of General Secretariat for Research & Technology (GSRT), under the Action "The Vineyard Roads" (project code: 2018ΣE01300000; Title of the project: Emblematic Research Action of National Scope for the exploitation of new technologies in the Agri-food sector, specializing in genomic technologies and pilot application in the value chains of "olive", "grapevine", "honey" and "livestock".

Acknowledgments: Authors would like to thank Interdisciplinary Agri-Food Center, Aristotle University of Thessaloniki (KEAGRO-AUTH), for providing access to the facilities of the unit.

Conflicts of Interest: The authors declare no conflict of interest.

References

1. Ramawat, K.G.; Mérillon, J.M. *Natural Products: Phytochemistry, Botany and Metabolism of Alkaloids, Phenolics and Terpenes*; Springer: Berlin/Heidelberg, Germany, 2013; pp. 1541–2662. [CrossRef]
2. Van Leeuwen, C.; Friant, P.; Chone, X.; Tregoat, O.; Koundouras, S.; Dubourdieu, D. Influence of Climate, Soil, and Cultivar on Terroir. *Am. J. Enol. Vitic.* **2004**, *55*, 207–217.
3. Kotseridis, Y.; Georgiadou, A.; Tikos, P.; Kallithraka, S.; Koundouras, S. Effects of Severity of Post-flowering Leaf Removal on Berry Growth and Composition of Three Red *Vitis vinifera* L. Cultivars Grown under Semiarid Conditions. *J. Agric. Food Chem.* **2012**, *60*, 6000–6010. [CrossRef] [PubMed]
4. Kallithraka, S.; Mohdaly, A.A.-A.; Makris, D.P.; Kefalas, P. Determination of major anthocyanin pigments in Hellenic native grape varieties (*Vitis vinifera* sp.): Association with antiradical activity. *J. Food Compos. Anal.* **2005**, *18*, 375–386. [CrossRef]
5. Monagas, M.; Gómez-Cordovés, C.; Bartolomé, B.; Laureano, O.; Ricardo Da Silva, J.M. Monomeric, Oligomeric, and Polymeric Flavan-3-ol Composition of Wines and Grapes from *Vitis vinifera* L. Cv. Graciano, Tempranillo, and Cabernet Sauvignon. *J. Agric. Food Chem.* **2003**, *51*, 6475–6481. [CrossRef]
6. Mattivi, F.; Scienza, A.; Failla, O.; Villa, P.; Anzani, R.; Tedesco, G.; Gianazza, E.; Righetti, P. Vitis vinifera-a chemotaxonomic approach: Anthocyanins in the skin. In Proceedings of the 5th International Symposium on Grape Breeding, Sankt Martin, Germany, 12–16 September 1989; pp. 119–133.
7. Liang, Z.; Wu, B.; Fan, P.; Yang, C.; Duan, W.; Zheng, X.; Liu, C.; Li, S. Anthocyanin composition and content in grape berry skin in Vitis germplasm. *Food Chem.* **2008**, *111*, 837–844. [CrossRef]
8. Garcia-Muñoz, S.; Muñoz-Organero, G.; De Andrés, M.T.; Cabello, F. Ampelography-An old technique with future uses: The case of minor varieties of *Vitis vinifera* L. from the Balearic Islands. *Oeno One* **2011**, *45*, 125–137. [CrossRef]
9. Gutiérrez-Gamboa, G.; Liu, S.Y.; Pszczólkowski, P. Resurgence of minority and autochthonous grapevine varieties in South America: A review of their oenological potential. *J. Sci. Food Agric.* **2020**, *100*, 465–482. [CrossRef]
10. Han, F.L.; Xu, Y. Effect of the Structure of Seven Anthocyanins on Self-Association and Colour in an Aqueous Alcohol Solution. *S. Afr. J. Enol. Vitic.* **2015**, *36*, 105–116. [CrossRef]
11. Manzano, S.G.; Santos-Buelga, C.; Dueñas, M.; Rivas-Gonzalo, J.-C.; Escribano-Bailón, T. Colour implications of self-association processes of wine anthocyanins. *Eur. Food Res. Technol.* **2008**, *226*, 483–490. [CrossRef]
12. Castellarin, S.D.; Di Gaspero, G. Transcriptional control of anthocyanin biosynthetic genes in extreme phenotypes for berry pigmentation of naturally occurring grapevines. *BMC Plant Biol.* **2007**, *7*, 46. [CrossRef]
13. Liu, Y.; Zhang, Y.; Zhou, Y.; Feng, X. Anthocyanins in Different Food Matrices: Recent Updates on Extraction, Purification and Analysis Techniques. *Crit. Rev. Anal. Chem.* **2022**, 1–32. [CrossRef] [PubMed]
14. Manzoor, M.F.; Hussain, A.; Naumovski, N.; Ranjha, M.M.A.N.; Ahmad, N.; Karrar, E.; Xu, B.; Ibrahim, S.A. A Narrative Review of Recent Advances in Rapid Assessment of Anthocyanins in Agricultural and Food Products. *Front. Nutr.* **2022**, *9*, 901342. [CrossRef] [PubMed]
15. López-fernández, O.; Domínguez, R.; Pateiro, M.; Munekata, P.E.S.; Rocchetti, G.; Lorenzo, J.M. Determination of Polyphenols Using Liquid Chromatography–Tandem Mass Spectrometry Technique (LC–MS/MS): A Review. *Antioxidants* **2020**, *9*, 479. [CrossRef] [PubMed]
16. Kalogiouri, N.P.; Samanidou, V.F. Liquid chromatographic methods coupled to chemometrics: A short review to present the key workflow for the investigation of wine phenolic composition as it is affected by environmental factors. *Environ. Sci. Pollut. Res.* **2021**, *28*, 59150–59164. [CrossRef]
17. Oancea, S.; Stoia, M.; Coman, D. Effects of Extraction Conditions on Bioactive Anthocyanin Content of Vaccinium Corymbosum in the Perspective of Food Applications. *Procedia Eng.* **2012**, *42*, 489–495. [CrossRef]
18. Chandra Singh, M.; Kelso, C.; Price, W.E.; Probst, Y. Validated liquid chromatography separation methods for identification and quantification of anthocyanins in fruit and vegetables: A systematic review. *Food Res. Int.* **2020**, *138*, 109754. [CrossRef]
19. Kallithraka, S.; Aliaj, L.; Makris, D.P.; Kefalas, P. Anthocyanin profiles of major red grape (*Vitis vinifera* L.) varieties cultivated in Greece and their relationship with in vitro antioxidant characteristics. *Int. J. Food Sci. Technol.* **2009**, *44*, 2385–2393. [CrossRef]
20. Kyraleou, M.; Kallithraka, S.; Gkanidi, E.; Koundouras, S.; Mannion, D.T.; Kilcawley, K.N. Discrimination of five Greek red grape varieties according to the anthocyanin and proanthocyanidin profiles of their skins and seeds. *J. Food Compos. Anal.* **2020**, *92*, 103547. [CrossRef]
21. Pinasseau, L.; Vallverdú-Queralt, A.; Verbaere, A.; Roques, M.; Meudec, E.; Le Cunff, L.; Péros, J.-P.; Ageorges, A.; Sommerer, N.; Boulet, J.-C.; et al. Cultivar Diversity of Grape Skin Polyphenol Composition and Changes in Response to Drought Investigated by LC-MS Based Metabolomics. *Front. Plant Sci.* **2017**, *8*, 1826. [CrossRef]
22. Kyraleou, M.; Koundouras, S.; Kallithraka, S.; Theodorou, N.; Proxenia, N.; Kotseridis, Y. Effect of irrigation regime on anthocyanin content and antioxidant activity of *Vitis vinifera* L. cv. Syrah grapes under semiarid conditions. *J. Sci. Food Agric.* **2016**, *96*, 988–996. [CrossRef]
23. Koundouras, S.; Kanakis, I.; Drossou, E.; Kallithraka, S.; Kotseridis, Y. Effects of postveraison water regime on the phenolic composition of grapes and wines of cv. Agiorgitiko (*Vitis vinifera* L.). *Oeno One* **2013**, *47*, 115–128. [CrossRef]
24. Chandra Singh, M.; Probst, Y.; Price, W.E.; Kelso, C. Relative comparisons of extraction methods and solvent composition for Australian blueberry anthocyanins. *J. Food Compos. Anal.* **2021**, *105*, 104232. [CrossRef]

25. Pourcel, L.; Irani, N.G.; Lu, Y.; Riedl, K.; Schwartz, S.; Grotewold, E. The Formation of Anthocyanic Vacuolar Inclusions in Arabidopsis thaliana and Implications for the Sequestration of Anthocyanin Pigments. *Mol. Plant* **2010**, *3*, 78–90. [CrossRef] [PubMed]
26. Barnes, J.S.; Nguyen, H.P.; Shen, S.; Schug, K.A. General method for extraction of blueberry anthocyanins and identification using high performance liquid chromatography–electrospray ionization-ion trap-time of flight-mass spectrometry. *J. Chromatogr. A* **2009**, *1216*, 4728–4735. [CrossRef]
27. Naczk, M.; Shahidi, F. Phenolics in cereals, fruits and vegetables: Occurrence, extraction and analysis. *J. Pharm. Biomed. Anal.* **2006**, *41*, 1523–1542. [CrossRef]
28. Czitrom, V. One-Factor-at-a-Time versus Designed Experiments. *Am. Stat.* **1999**, *53*, 126–131. [CrossRef]
29. Paun, N.; Botoran, O.R.; Niculescu, V.-C. Total Phenolic, Anthocyanins HPLC-DAD-MS Determination and Antioxidant Capacity in Black Grape Skins and Blackberries: A Comparative Study. *Appl. Sci.* **2022**, *12*, 936. [CrossRef]
30. Trikas, E.D.; Papi, R.M.; Kyriakidis, D.A.; Zachariadis, G.A. A Sensitive LC-MS Method for Anthocyanins and Comparison of Byproducts and Equivalent Wine Content. *Separations* **2016**, *3*, 18. [CrossRef]
31. Mattivi, F.; Guzzon, R.; Vrhovsek, U.; Marco, S.; Velasco, R. Metabolite profiling of grape: Flavonols and anthocyanins. *J. Agric. Food Chem.* **2006**, *54*, 7692–7702. [CrossRef]
32. Cosme, F.; Pinto, T.; Vilela, A. Phenolic Compounds and Antioxidant Activity in Grape Juices: A Chemical and Sensory View. *Beverages* **2018**, *4*, 22. [CrossRef]
33. Kallithraka, S.; Tsoutsouras, E.; Tzourou, E.; Lanaridis, P. Principal phenolic compounds in Greek red wines. *Food Chem.* **2006**, *99*, 784–793. [CrossRef]
34. Costa, E.; Cosme, F.; Jordão, A.M.; Mendes-Faia, A. Anthocyanin profile and antioxidant activity from 24 grape varieties cultivated in two Portuguese wine regions. *Oeno One* **2014**, *48*, 51–62. [CrossRef]
35. Zhao, Q.; Duan, C.-Q.; Wang, J. Anthocyanins Profile of Grape Berries of Vitis amurensis, Its Hybrids and Their Wines. *Int. J. Mol. Sci.* **2010**, *11*, 2212–2218. [CrossRef] [PubMed]
36. Arnous, A.; Makris, D.P.; Kefalas, P. Anthocyanin Composition and Colour Characteristics of Selected Aged Wines Produced in Greece. *J. Wine Res.* **2002**, *13*, 23–34. [CrossRef]
37. Biniari, K.; Xenaki, M.; Daskalakis, I.; Rusjan, D.; Bouza, D.; Stavrakaki, M. Polyphenolic compounds and antioxidants of skin and berry grapes of Greek Vitis vinifera cultivars in relation to climate conditions. *Food Chem.* **2020**, *307*, 125518. [CrossRef] [PubMed]
38. García-Beneytez, E.; Revilla, E.; Cabello, F. Anthocyanin pattern of several red grape cultivars and wines made from them. *Eur. Food Res. Technol.* **2002**, *215*, 32–37. [CrossRef]
39. Kallithraka, S.; Arvanitoyannis, I.S.; Kefalas, P.; El-Zajouli, A.; Soufleros, E.; Psarra, E. Instrumental and sensory analysis of Greek wines; implementation of principal component analysis (PCA) for classification according to geographical origin. *Food Chem.* **2001**, *73*, 501–514. [CrossRef]
40. Ribéreau-Gayon, P.; Glories, Y.; Maujean, A.; Dubourdieu, D. *Handbook of Enology, The Chemistry of Wine: Stabilization and Treatments*, 2nd ed.; Wiley: Hoboken, NJ, USA, 2006; Volume 2, ISBN 9780470010396.
41. Baiano, A.; Terracone, C. Varietal Differences among the Phenolic Profiles and Antioxidant Activities of Seven Table Grape Cultivars Grown in the South of Italy Based on Chemometrics. *J. Agric. Food Chem.* **2011**, *59*, 9815–9826. [CrossRef]
42. Philippidis, A.; Poulakis, E.; Basalekou, M.; Strataridaki, A.; Kallithraka, S.; Velegrakis, M. Characterization of Greek Wines by Ultraviolet–Visible Absorption Spectroscopy and Statistical Multivariate Methods. *Anal. Lett.* **2017**, *50*, 1950–1963. [CrossRef]
43. Ma, T.-T.; Sun, X.-Y.; Gao, G.-T.; Wang, X.-Y.; Liu, X.-Y.; Du, G.-R.; Zhan, J.-C. Phenolic Characterisation and Antioxidant Capacity of Young Wines Made From Different Grape Varieties Grown in Helanshan Donglu Wine Zone (China). *S. Afr. J. Enol. Vitic.* **2014**, *35*, 321–331. [CrossRef]
44. *SANTE/12682/2019*; Guidance Document on Analytical Quality Control and Method Validation for Pesticide Residues Analysis in Food and Feed. Safe Food Chain Pesticides Biocides European Commission, 2019; pp. 1–48.
45. Koskela, A.K.J.; Anttonen, M.J.; Soininen, T.H.; Saviranta, N.M.M.; Auriola, S.; Julkunen-Tiitto, R.; Karjalainen, R.O. Variation in the Anthocyanin Concentration of Wild Populations of Crowberries (*Empetrum nigrum* L subsp. *hermaphroditum*). *J. Agric. Food Chem.* **2010**, *58*, 12286–12291. [CrossRef] [PubMed]

molecules

MDPI

Review

Recent Advances in Mycotoxin Determination in Fish Feed Ingredients

Sofia Vardali [1,*], Christina Papadouli [1], George Rigos [2], Ioannis Nengas [2], Panagiota Panagiotaki [1] and Eleni Golomazou [1,*]

[1] Department of Ichthyology and Aquatic Environment—Aquaculture Laboratory, School of Agricultural Sciences, University of Thessaly, 38446 Volos, Greece
[2] Institute of Marine Biology, Biotechnology and Aquaculture, Hellenic Centre for Marine Research, 46.7 km Athens-Sounion, 19013 Attiki, Greece
* Correspondence: sofvardali@gmail.com (S.V.); egolom@uth.gr (E.G.)

Abstract: Low-cost plant-based sources used in aquaculture diets are prone to the occurrence of animal feed contaminants, which may in certain conditions affect the quality and safety of aquafeeds. Mycotoxins, a toxic group of small organic molecules produced by fungi, comprise a frequently occurring plant-based feed contaminant in aquafeeds. Mycotoxin contamination can potentially cause significant mortality, reduced productivity, and higher disease susceptibility; thus, its timely detection is crucial to the aquaculture industry. The present review summarizes the methodological advances, developed mainly during the past decade, related to mycotoxin detection in aquafeed ingredients, namely analytical, chromatographic, and immunological methodologies, as well as the use of biosensors and spectroscopic methods which are becoming more prevalent. Rapid and accurate mycotoxin detection is and will continue to be crucial to the food industry, animal production, and the environment, resulting in further improvements and developments in mycotoxin detection techniques.

Keywords: mycotoxins; aquafeeds; aquaculture; cereals; detection; HPLC; ELISA; FT-NIR; biosensors

1. Introduction

The aquaculture industry has become a source of high-quality protein providers for humans worldwide [1]. It is one of the fastest-growing industries in food production with an average growth rate of 5.3% between 2001 and 2018 [1], globally accounting for more fish biomass than capture fisheries [2]. Fish stock depletion, rapid increase in global population, high demand for seafood products, and international trade have contributed to the tremendous aquaculture expansion during the past decades [3]. The challenge of the aquaculture industry to meet increasing fish demand and achieve food security goals within environmental boundaries will become critical in the coming years, especially with a global population headed to 10 billion by 2050 [4,5].

Marine ingredients are highly important in aquatic feed to provide macro- and micronutrients and organoleptic properties, enhancing the digestibility and growth performance of formulated diets. Marine ingredients used in aquafeed are usually meals and oils rendered by small pelagic fish, and by-products of fish and seafood processing [6]. Fish meal is considered the most valuable protein source, due to its exceptional benefits including its well-balanced composition of amino acids, good digestibility, and palatability, as well as its enhancement of the digestion, and absorption of nutrients in fish diets [7].

About 70% of aquaculture production depends on providing aquatic animals with high-quality, rich in protein aquafeeds [1]. Aquaculture globally is currently consuming around 69% of fishmeal and 75% of fish oil supplies [8] and as it is continuously expanding, the demand for fishmeal and fish oil produced by marine pelagic fisheries will be steadily increasing. This has led to a progressive decline in these fish stocks and caused severe inflation in their global prices [9]. Given the increased demand for high-quality fish species, aquaculture addresses the challenge of succeeding in sustainable growth by replacing fish protein sources with plant and terrestrial animal proteins, without compromising the economic value and quality of the final product [10].

Plant-based protein sources have been generally used to partially substitute fish meal in fish diets [11]. These protein sources are advantageous by having a high content of available protein, continuous availability, environmental sustainability, and affordable prices [12–14]. Plant-based feed ingredients also provide dietary carbohydrates which can be a source of energy for fish and shrimp; depending on their ability to utilize dietary carbohydrates for energy depending on the species and their natural diet [15]. Furthermore, plant ingredients are sources of starch which is necessary as a binder and facilitates extruded pellet expansion [16].

Plant-based feed ingredients currently used in aquafeeds as substitutes for marine ingredients include among others, soybean meal, rapeseed/canola meal, maize/corn, wheat bran, wheat, and barley [17]. Corn gluten is also a promising ingredient in fish feed, due to its high nutrient content and its increased availability as a bioethanol production by-product [18]. Corn gluten and wheat gluten are high in protein, low in fiber, rich in vitamins B and E, and do not contain any antinutritional factors [19]. Sunflower meal is highly palatable and has low antinutritional factors [20]. Soybean meal is one of the most interesting alternatives to fishmeal because of the advantages of easy supply, low price, and increased protein and amino acid composition [21]. However, soybean meal has been found to induce a variety of histological and functional changes in the gastrointestinal tracts of several species, such as subacute enteritis of the distal epithelial mucosa including morphological alteration and inflammation [22].

As opposed to the aforementioned benefits of plant-based sources, animal diets based on plant proteins can be often associated with reduced feed intake, growth performance, and intestinal function [13,22,23]. Some ingredients of plant origin have certain characteristics, such as high carbohydrate content, deficiency in some essential amino acids, low palatability, as well as content in some anti-nutritional factors [24], that limit their use. Both the quality and safety of aquaculture diets can be affected by anti-nutritional compounds including phytates, protease inhibitors, saponins, glycosylates, and tannins, resulting from the inclusion of plant dietary sources [25]. Fish diets may be also contaminated by mycotoxins, also derived from plant-based raw materials [26].

Mycotoxins are secondary metabolites produced by various species of fungi, often found in agricultural products that are used to feed livestock. These toxins pose a health risk to both livestock and consumers. Agricultural raw materials can be contaminated by fungi during the growing process, before harvest, or during storage in inadequate conditions of humidity and/or temperature [25]. Since mycotoxins are natural contaminants and pose a health risk to both livestock and consumers, several European and international organizations have dealt with this issue by identifying their particular importance and establishing regulatory limits and proposing recommended levels for selected mycotoxins. These include the European Commission (EC, Brussels, Belgium), the US Food and Drug Administration (FDA, Silver Spring, MD, USA), the Food and Agriculture Organization of the United Nations (FAO, Rome, Italy), and the World Health Organization (WHO, Geneva, Switzerland). A scientific expert committee jointly convened by WHO and FAO, named JECFA, serves actually as the international body responsible for evaluating the health risk from all natural toxins including mycotoxins.

Several mycotoxins have been identified and those of significant importance in animal feeds are primarily produced by the five fungal genera *Aspergillus*, *Fusarium*, *Penicillium*, *Claviceps*, and *Alternaria* [27]. Approximately 400 compounds have been identified as mycotoxins [28]. Mycotoxins may cause health issues in livestock when accidentally present alone or synergistically in animal diets [29]. The aflatoxins (AFs) such as aflatoxin B1, B2, G1, G2, and M1 are human and animal health hazards according to the International Agency for Research on Cancer (IARC) [30]. Moreover, ochratoxins (OTA), and fumonisins (FBs) B1,

B2, and B3 have been assessed as possible human carcinogens [31,32]. Furthermore, other mycotoxins have been considered as serious threats including trichothecenes (TCs) type A (HT-2 toxin and T-2 toxin) and B (deoxynivalenol-DON), zearalenone (ZEN), Fusarium mycotoxins, ergot alkaloids (EAs), Alternaria toxins (ATs) and patulin (PAT) [33].

AFB1, DON, ZEN, and FB1 belong to the most contaminants of animal feeds [34]. Mycotoxic contamination may considerably affect animal health, causing functional abnormalities, toxicity hepatic problems, immunotoxicity issues, and reduced growth and animal productivity [35–39]. ZEN may induce reproductive problems such as hyperestrogenism, sterility, and even abortions, affected by the estrogenic activity of ZEN which interferes with animal reproduction [40]. In fish feed, FBs and DON are among the most frequently detected mycotoxins at high levels [26]. These mycotoxins can potentially cause problems in fish farm operations, with significant economic losses such as mortality, reduced productivity, and higher susceptibility to diseases [39].

AFB1 is the only mycotoxin regulated by Directive 2002/32/EC of the European Parliament and of the Council of 7 May 2002 on undesirable substances. The maximum allowed concentration in feed materials for fish species is 20 μg/kg (ppb), and for complete feed is 10 ppb [41]. Concerning other important mycotoxins such as DON, ZEN, T-2 and HT-2 toxin, FB1 and FB2, the European Commission (EC) has established only recommended limits for their presence in feedstuffs and feed [42–44]. Among these recommended limits, only values for FB1 and FB2 refer directly to fish species. The recommended maximum concentration for DON is 8000 ppb for cereal and cereal products except for maize by-products, while for complementary and complete feeding stuffs the limit is 5000 ppb. For ZEN, the recommended limits are 2000 ppb for cereals and cereal products except for maize by-products. For OTA, the limit is 250 ppb for both cereals and cereals products. For the summary of FB1 and FB2, EC has proposed the limit of 10,000 ppb for complementary and complete feeding stuffs for fish. For T-2 and HT-2 toxins in cereals and cereal products, except for oat bran, the recommended limit is 500 ppb, while for ergot alkaloids found in feed containing unground cereals, the limit is 1000 ppb.

The potential hazards of mycotoxin presence in feed materials have driven efforts to develop various analytical methods for the identification and quantification of mycotoxins in food samples. Continuous improvements in mycotoxin analytical methodology using advanced and rapid techniques are paramount to comply with the updated legislation and protect consumers of aquatic products. Recently, the existed methodologies for myco-toxin detection related to human and animal health were reviewed in different food matrices [45,46]. The present review focuses primarily on the advances in mycotoxin detection during the last decade in plant-derived raw materials comprising the major fish feed ingredients. Extraction and analytical methods are briefly covered as well as consideration for the future of mycotoxin analysis.

2. Sampling and Sample Preparation Methods

2.1. Sampling and Sample Preparation

In the mycotoxin sampling process, it is essential to ensure accuracy and representativeness in the sample collection. Mycotoxins can be found in marginally detectable amounts, and products are not uniformly contaminated, increasing the risk of inaccurate sampling [47,48]. A specific protocol is followed to collect representative laboratory samples from all sampling points. The lots selected for inspection must be appropriate. The final sample is the combination of several replicate samples from different parts of the lot. This is achieved by mixing and dividing to obtain a representative sample [49,50].

Mycotoxin detection is typically done by collecting and testing samples of food or feed. The samples should be taken randomly from the entire lot and should be collected from different locations within the lot [50]. The sample size needed depends on the size of the lot, but a minimum of 500 g should be collected using clean and properly sterilized sampling equipment to avoid contamination [51]. Samples should be stored in airtight containers, at the appropriate temperature and humidity, and transported to the laboratory as quickly as possible to prevent degradation of the mycotoxin [51]. The goal of sample preparation is to create a representative and clean sample that can be accurately analyzed by the chosen analytical technique. The specific steps involved in sample preparation depend on the type of sample and the analytical technique used. To obtain an analytical part (test portion), a sample must be ground, homogenized, and subsampled. This analytical portion is then extracted with a solvent, analyzed, and the mycotoxin concentration is determined using a validated analytical approach [52].

Sample preparation is an important step in analytical techniques because it could greatly affect the accuracy and precision of the results.

2.2. Sampling Error

Sampling errors can occur in several ways and can greatly affect the accuracy and precision of analytical results. Sampling errors in mycotoxin detection refer to the potential for variability in the results of mycotoxin testing due to the collection and preparation of the sample [53]. Such errors may also occur when a sample is not representative of the entire lot, or when the sample is contaminated during collection or handling [54].

To minimize the sampling error, it is important to follow proper sampling procedures, including the selection of multiple samples from different locations within the batch and using appropriate sampling tools and techniques [55]. Additionally, samples should be properly stored and prepared according to established protocols. Quality control measures should be also implemented to ensure the accuracy and reliability of the testing results [56].

2.3. Sample Pre-Treatment (Extraction and/or Clean-Up)

2.3.1. Solid–Liquid Extraction (SLE)

Solid–liquid extraction involves a simple and high-sensitivity process for preparing samples in liquid chromatography–mass spectrometry (LC-MS). The multi-residue analysis ability of the detection apparatus is rather advantageous [50,57]. The capacity of matrix effects in LC-MS analysis to modify chromatographic signals is accomplished due to co-eluting matrix components, while the ion suppression problem that arises can be overcome by proper sample preparation using specific matrices and internal standards [57]. Apart from being simple and sensitive, sample preparation in this procedure is considered reliable and has been successfully applied to detect mycotoxins in plant samples. Few methods were found in pertinent literature using simple SLE extraction without any further clean-up steps. All methods used a solvent mixture of acetonitrile/water for the extraction of mycotoxins before LC-MS/MS analysis [58–61]. However, the simplicity of sample preparation process may affect the method's performance characteristics (low recoveries, matrix effect, etc.,) [62–64].

2.3.2. Dispersive-Solid Phase Extraction (d-SPE)

In the d-SPE technique, sample cleaning is achieved by the use of a solid sorbent in a liquid or dissolved sample which retains impurities. After separation, the sample is centrifuged for sorbent removal (Anand and Srivastava, 2020). Different types of sorbents are used. C18 sorbent is used to extract non-polar or relatively polar compounds, retaining most of the organic compounds present in an aqueous phase.

The QuEChERS protocol as named by quick, easy, cheap, effective, rugged, and safe, is a commonly applied extraction method that requires small amounts of sample and solvent while at the same time producing high extraction efficiency. This advantage explains its high popularity in current extraction techniques [55]. The extraction phase is the first step where an organic solvent such as acetonitrile is needed along with a variety of salts to modulate polarity and pH and to facilitate phase separation and recovery of the analyte. Purification is the second stage in the cleaning process. The remaining water and other interfering substances from the matrix are eliminated in this stage [65]. There are several methods using the QuEChERS protocols for the extraction and clean-up of mycotoxins before instrumental analysis [66–72].

2.3.3. Clean Up by Immunoaffinity Column (IAC)

The creation of IAC is another technique where specific antibodies for certain mycotoxins are bound to a specifically activated SP support. This method is commonly used to detect Afs, OTA, and FBs. In detail, the support is packed into a cartridge while a suspension is performed in a buffer solution. After the extract or fluid's mycotoxin attaches to the antibody and any contaminants are washed away with water or an aqueous solution, the mycotoxin is desorbed using a miscible solvent (methanol). IAC can be used for further separation and LC quantification [73]. A few methods were found in the literature using the IAC columns for mycotoxins clean-up step [74–77].

2.3.4. Solid Phase Extraction (SPE)

The SPE reduces matrix-based interferents to concentrate on a target analyte. To properly separate the analytes from the other interferents, this absorbent is selected based on the physicochemical characteristics of the analytes [78]. Dispersive SPE, is a modern technique that requires nanoparticles in a magnetic mode. This detection method has been recently adopted [79]. Notably, the characteristics of the magnetic SPE (mSPE) resemble those found in standard SPE. Matrix composition may affect the selection of the adsorbent and elution mixture [79]. In mSPE, continuous contact with the adsorbent is necessary through the dispersion of the magnetic material into the solution containing the target molecules.

A method using modified magnetic nanoparticles as a solid phase adsorbent for extraction of OTA in rice, wheat, and corn has been developed and very low limits of detection were achieved (0.03–0.06 μg/kg) while the recoveries are 87 to 93% [80]. Furthermore, another method using mSPE as a clean-up step has been developed for the determination of AFB1, AFB2, AFG1, AFG2, OTA, and ZEN in various kinds of cereal using a LC-MS/MS system [81].

2.3.5. Molecular Imprinted Polymer (MIP)

For clean-up and preconcentration of mycotoxins, a new class of intelligent polymers based on MIPs has proven to be an effective technique. The MIP is a synthetic material with an artificially generated three-dimensional network that can specifically rebind a target molecule. MIP is cost-effective, chemically, and thermally stable and compatible with all solvents [82]. A magnetic MIP (mMIP) with quercetin as a dummy template has been used for the extraction of ZEN from maize, wheat, and rice by Cavaliere et al. 2019 [83].

2.3.6. Ultrasonic Solvent Extraction (USE)

The USE method involves mechanical wave propagation that makes up an ultrasound created by cycles of compression and refraction, or waves with high and low pressures combined in frequencies above 20 kHz. Temperature and pressure changes may affect USE operation by the creation of bubbles. Both particle collisions and ultrasonic waves can cause fragmentation, which decreases particle size and aids in mass transfer [84]. The combination of USE with SLE extraction is used in an LC-MS technique for mycotoxin detection [75].

3. Instrumental Analysis

3.1. Chromatographic Methods—Detection Systems

Chromatographic-based methods include liquid chromatography (LC) or gas chromatography (GC) that is coupled with ultraviolet (UV), mass spectrometry (MS), or fluorescence (FLD). The chromatographic methods combined with a UV detector and FLD are usually used for the analysis of a compound or a small number of mycotoxin-related chemicals. The MS method has many advantages such as high sensitivity, selectivity, and accuracy, compared to the two other methods. Tandem MS (MS/MS), where two MS equipment are coupled together, is a highly sensitive, specific, and reliable tool for detecting contaminants in food and has become the most popular approach for multianalyte analyses [85,86]. LC-tandem MS (LC-MS/MS) has been increasingly used for the accurate quantitative analysis of mycotoxins in food [87]. A limited number of multi-mycotoxin techniques, particularly for finished fish feeds and shrimp feeds, has been reported. Aquatic feeds are complex matrices consisting of minerals, vitamins, fatty acids, and proteins in high concentrations that are challenging to remove [88]. As a result, choosing an appropriate clean-up step that reduces matrix effects (MEs) and interferences during chromatographic analysis is essential [89]. The literature on exclusive analyzes of fish feed ingredients seems insufficient, therefore, the data are mainly based on analyses of raw materials used in all types of animal feed, including fish feed. Table 1 summarizes the methods reviewed mainly in the past decade using HPLC systems equipped with several detection systems for the determination of mycotoxins in fish feed ingredients and aquafeeds.

Table 1. HPLC methods used in the analysis of mycotoxins.

Type of Cereal	Mycotoxins	Extraction Process—Clean-up	Analytical Technique	Recovery %	Limit of Detection (LOD)	Ref.
Maize, Wheat, Barley	11 mycotoxins:	SLE acetonitrile/water/formic acid (79/20/1, *v/v/v*)	UHPLC-MS/MS	63.2–111.2%	0.15–61 μg/kg	[59]
Barley, Wheat, Oat	10 mycotoxins	SLE: 84% (*v/v*) aqueous acetonitrile with 1% (*v/v*) formic acid Clean-up: d-SPE (mixture octadecyl silica and primary-secondary amine)	UPLC-MS/MS	83.3–92.8%	0.13–3.56 μg/kg	[90]
Barley, Wheat, Oat	23 mycotoxins	QuEChERS Extraction: Acetonitrile 5% formic acid Clean-up: QuEChERS ($MgSO_4$ and NaCl)	LC-MS/MS	70.1–109.3%	0.03–2.17 μg/kg	[69]
Corn, Oat	T-2 and HT-2	Extraction: ethanol-water (80:20; *v/v*)- Clean-up: IAC	UPLC-MS/MS	78.6–98.6 %	0.02–0.08 μg/kg	[74]
Corn	ZEN, α-zearalenol (α-ZEL), β-zearalenol (β-ZEL), α-zearalanol (α-ZAL), β-zearalanol (β-ZAL), zearalanone (ZAN)	Extraction: acetonitrile/water (90/10; *v/v*). Clean-up: SPE using a MycoSep 226 column	Isotope dilution-liquid chromatography/tandem mass spectrometry (ID-UPLC-MS/MS)	96.7−103.6%.	0.14–0.33 μg/kg	[91]
Corn, wheat	T-2, HT-2, diacetoxyscirpenol (DAS) and neosolaniol (NEO)	Extraction: acetonitrile/water, 84/16; (*v/v*) Clean-up: SPE withMycoSep 227 column	ID-UPLC-MS/MS	97–103%	0.01–0.12 μg/kg	[92]
Maize, Oat	DON and T-2	Extraction: acetonitrile/water mixture Clean-up: SPE by MycoSep 227 columns	UPLC-MS/MS	85.0–95.3%	0.13–0.38 μg/kg	[93]
Wheat, Corn Rice, Barley	38 (modified) mycotoxins	QuEChERS Extraction: acetonitrile/water/formic acid (75:20:5, *v/v/v*) d-SPE: anhydrous $MgSO_4$, NaCl, Na_2H-citrate·$1.5H_2O$, Na_3-citrate·$2H_2O$	LC-MS/MS	61−120%	LOQ: 0.05−80.0 μg/kg for wheat, 0.07−120 μg/kg for corn, 0.05−150 μg/kg for rice, and 0.10−150 μg/kg for barley	[70]
Maize, Wheat, Rice	ZEN	Extraction: acetonitrile/water, 80:20 (*v/v*) with 0.2% HCOOH Clean-up: mMIPs	UHPLC-MS/MS	>95%	0.044 μg/kg	[83]
Maize	AFB1, AFB2, AFG1, AFG2, OTA, ZEN, T2, FB1, FB2	Extraction: 2 SLE steps with acetonitrile 80% (*v/v*)	UHPLC-ToF-MS	77.8–110.4%	0.5–62.5 μg/kg	[61]
Corn meal, Durum, wheat flour	AFB1, AFB2, AFG1, AFG2, OTA, ZEN	Extraction: acetonitrile/water/formic acid 80:19.8:0.2 (*v/v/v*) Clean-up: mSPE	LS-MS/MS	>60%	0.05–2.2 μg/kg	[81]
Wheat flours, Corn meal and other cereal-derived products	AFB1, AFB2, AFG1, AFG2, T-2, HT, FB1, FB2	QuEChERS Extraction: H_2O 0.1% formic acid, Clean-up: Acetonitrile d-SPE: $MgSO_4$ and NaCl	LC-MS/MS	83.6–102.9%	0.5–100 μg/kg	[66]
Maize, Wheat, Sunflower, Soybean, Barley, Feeds, Feedstuffs	22 mycotoxins	QuEChERS Extraction: 2% acetic acid solution, Clean-up: Acetonitrile d-SPE: $MgSO_4$ and NaCl	UHPLC-MS/MS	67–94%	0.064–119.04 μg/kg	[68]
Maize, Wheat	11 mycotoxins	SLE extraction: acetonitrile/water mixture	UPLC-MS/MS	52.8–113.9%.	0.08–30.0 μg/kg	[58]
Maize, Rice	10 mycotoxins	USE extraction after the addition of $MeOH/H_2O/CHCl_3$ (75:20:5, *v/v/v*) and NaCl	LC-MS second-order calibration method based on alternating trilinear decomposition (ATLD) algorithm	93.8–109%	0.01–1.17 μg/kg	[94]

Table 1. *Cont.*

Type of Cereal	Mycotoxins	Extraction Process—Clean-up	Analytical Technique	Recovery %	Limit of Detection (LOD)	Ref.
Wheat	10 mycotoxins	QuEChERS Extraction (acetonitrile–water (84/16)) d-SPE: QuEChERS (PSA and C18)	UHPLC-MS/MS	70–116%	LOQ < 7 μg/kg	[72]
Maize, Oat, Rice, Rye, Barley, Wheat	AFB1, AFB2, AFG1, AFG2, DON	Extraction: 2 extractions with water and a mixture of methanol/water clean-up: SPE	HPLC–DAD–FLD	90–112%	0.02–16.2 μg/kg	[95]
Corn, Wheat, Barley	20 Fusarium toxins	Extraction: 2% acetic acid aqueous solution/acetonitrile (1:1, *v*/*v*) clean-up: QuEChERS	LC-Orbitrap MS	71–106%	LOQ: 5 μg/kg	[71]
Barley, Malt	17 mycotoxins	Extraction: (0.1% HCOOH/cetonitrile (1:1, *v*/*v*) Clean-up: QuEChERS ($MgSO_4$ + NaCl)	UPLC-MS/MS	75–124% Except of Nivalenol 50–51%	0.3–24 μg/kg	[67]
Rice, Wheat, Corn	OTA	Extraction: SLE Clean-up: mSPE	LC-FLD	87–93%	0.03–0.06 μg/kg	[80]
Rice, Wheat, Oat, Maize, Barley	11 mycotoxins	SLE extraction acetonitrile: water: acetic acid, 79:20:1	UPLC-MS/MS	83.5–107.3%	0.01–25 μg/kg	[86]
Oats, Wheat	HT-2 and T-2 toxins	Extraction: methanol/water (90:10, *v*/*v*) clean-up: immunoaffinity columns	UPLC-PDA	87–103%	8 μg/kg	[76]
Maize, Wheat, Oats, Cornflakes, Bread	14 mycotoxins	Extraction: acetonitrile/water/acetic acid (79/20/1, *v*/*v*/*v*) followed by a hexane defatting step	LC-MS/MS	70–110%	5–13 μg/kg	[77]
Wheat	OTA	Extraction: methanol/3% aqueous sodium bicarbonate (3/7, *v*/*v*) Clean-up: MIP spe column	Automated SPE system with on-line fluorescence detection MISPE-FLD	84–102%	1.2 ng/mL	[96]
Maize, Oats	DON and T-2	Extraction: acetonitrile/water (84:16; *v*/*v*) Clean-up: SPE column	UPLC-MS/MS	85.0–95.3%	0.04–0.12 μg/kg	[97]
Wheat, Maize	35 mycotoxins	QuEChERS: extraction/partition process) of 5% formic acid in acetonitile ($MgSO_4$ and NaCl)	UPLC-MS/MS	60–103%	0.13–23.99 μg/kg	[98]
Wheat, Corn, Oat, Barley, Rice	AFB1, AFB2, AFG1, AFG2, OTA, and ZEN	Extraction: 80% methanol Clean-up: multifunctional immunoaffinity column	HPLC-FLD Using a photochemical reactor enhance derivatization system (PHRED)	77–104%	0.004–0.5 μg/kg	[99]
Barley, Oat, Wheat	16 mycotoxins	Extraction: SLE acetonitrile:water:acetic acid (79:20:1, *v*/*v*/*v*)	LC-MS/MS	84–116%	0.1–4.3 μg/kg	[60]
Fish feed and shrimp feed	AFB1, AFM1, T-2, HT-2, DON, OTA, and ZEN	acetonitrile–water (3 + 1, *v*/*v*) saturated hexane clean-up by multitoxin column	HPLC-MS/MS	80.5 to 116.5%	1.83–12.63 lg/kg	[88]
Fish feed	DON and ZEN	SLE Clean-up: IAC column	HPLC-DAD	79–90%	2–30 μg/kg	[100]
Fish feed	15 mycotoxins	USE extraction Clean-up: Captiva EMR Lipid cartridge	LC-MS/MS	25–109%	0.05–54 μg/kg	[101]

Most of the methods found in the literature use LC-MS/MS as a detection system. Some of them using various extraction processes before LC-MS/MS analysis for the mycotoxin determination are detailed below. LC-MS/MS method was reported for the analysis of 15 mycotoxins in fish feed. The extraction was achieved by a USE step followed by a clean-up step by a lipid cartridge. The recoveries varied between 25 and 109% for the 15 mycotoxins and LODs ranged between 0.05 and 54 μg/kg [101]. Furthermore, for the detection of AFB1, AFM1, T-2, HT-2, DON, OTA, and ZEN in fish feed and shrimp feed, an HPLC-MS/MS method was developed. The samples were extracted with a mixture of acetonitrile and water followed by a defatted step by hexane and a clean-up step with a multi-toxin purification column. LODs ranged between 1.83 and 12.63 μg/kg and the method was successfully applied in several fish feeds in China [88]. The LODs of the method ranged between 0.15 and 61 μg/kg [59]. Another method for the simultaneous determination of ZEN and DON in fish feed was reported based on an HPLC-DAD system and an SLE extraction step followed by clean-up with an IAC column [100].

Concerning fish feed ingredients, the simultaneous determination of 11 mycotoxins in maize, wheat, and barley was achieved by UHPLC-MS/MS analysis using a simple SLE extraction with acetonitrile/water/formic acid (79/20/1, *v*/*v*/*v*). The UPLC-MS/MS method for the analysis of DON and T-2 in maize and oats was also developed, followed by SLE extraction and an SPE clean-up step. The LODs of the method were between 0.13 and 0.38 μg/kg [93]. In wheat, corn, rice, and barley, LC-MS/MS was applied for the determination of 38 mycotoxins using a QuEChERS extraction [70]. Another LC-MS/MS method was developed for the detection of T-2 and HT-2 toxins in corn and oat using an SLE extraction followed by a clean-up step with an IAC column, achieving very low LODs, ranging between 0.02 and 0.08 μg/kg [74].

Only four methods were found using an FLD detection system. In rice, wheat, and corn samples, the mSPE extraction before LC-FLD detection was applied for the analysis of OTA with LODs ranging from 0.03 to 0.06 μg/kg [80]. For the detection of aflatoxins AFB1, AFB2, AFG1, AFG2, and DON in maize, oat, rice, rye, barley, and wheat an HPLC-DAD-FLD system was developed, after two SLE extractions and a clean-up step by SPE. The LODs of the method were 0.02 to 16.2 μg/kg [95]. An automated molecularly imprinted SPE system with online fluorescence detection MISPE-FLD was applied for the determination of OTA in wheat samples achieving LOD 1.2 ng/mL [96]. In wheat, corn, oat, barley, and rice, a validated HPLC-FLD system coupled with a photochemical reactor was tested for the simultaneous determination of aflatoxins, OTA, and ZEN. The LODs of the method ranged between 0.004 and 0.5 μg/kg [99]. Full scan MS was also found in pertinent literature using a TOF-MS and an Orbitrap MS system for the identification of mycotoxins [61,71,94]. Orbitrap MS and TOF MS are used to estimate both known and unknown compounds. This is because they have the ability to allow detailed discrimination in molecular weight by accurately measuring the mass to five significant digits [71]. A validated UHPLC-ToF-MS method was developed for the determination of nine mycotoxins in maize. The extraction step proposed was very easy, using two SLE steps with acetonitrile 80% (*v*/*v*). The method's LODs ranged between 0.5 and 62.5 μg/kg [61]. Furthermore, the LC-Orbitrap MS method combined with QuEChERS step was applied for the determination of 20 fusarium toxins in corn, wheat, barley, sunflower, soybean, feeds, and feedstuffs and the LODs of the method were 5 μg/kg [68]. Ten mycotoxins in maize and rice were detected by a full-scan LC-MS method using a second-order calibration method based on an alternating trilinear decomposition (ATLD) algorithm. The extraction was achieved using a USE extraction after the addition of $MeOH/H_2O/CHCl_3$ (75:20:5, *v*/*v*/*v*) and NaCl. The LODs of the method ranged between 0.01 and 1.17 μg/kg [94].

A UPLC method coupled with a photo diode array detector (DAD) for the analysis of T-2 and HT-2 toxins in oats and wheat has been also evaluated. The extraction solvents used were methanol/water (90:10, *v*/*v*) followed by a clean-up step with an IAC column. The LOD of the method for the two toxins was 8 μg/kg [76].

3.2. *Immunological Methods (Enzyme-Linked Immunosorbent Assay-ELISA)*

Lateral flow immunoassay, ELISA, and immunosensors are immunochemical detection methods based principally on antibody–antigen binding [102]. Antibodies and antigens belong to some of the most commonly used capture agents in immunoassays for disease treatment, environmental monitoring, and food safety regulation. Their high commercial recognition is, however, not deprived of drawbacks. For example, immunization and purification are necessary for the development of high-quality antibodies. These processes can be difficult, expensive, and laborious. Additionally, the applicability of antibodies is limited due to their sensitivity to pH and temperature variation. Moreover, antibodies can only recognize substances that are immunogenic and immunoreactive. Finally, the chemical conjugation effectiveness of mycotoxins to a protein carrier is limited [103,104].

Immunosorbent assays and immunosensors require simpler sample pre-treatment compared to those required for chromatographic methods and have the advantages of high throughput and good specificity although, detection results still must be output by instruments. Under the same sample pre-treatment procedure, ELISA assays are however more prone to more errors due to the tedious operation process [105]. Therefore, developing a more sensitive and rapid on-site detection assay is urgently needed to detect toxic and harmful substances in food. Several studies using a variety of assays and detectors have been reported in the literature and are presented in Table 2.

Table 2. ELISA methods used in the analysis of mycotoxins.

Type of Cereal	Mycotoxins	Method	Detection Method	LOD	Ref.
Maize, wheat, vegetable oil samples	ZEN	Fluorescence quenchometric lateral flow immunochromatographic assay	UV-absorbance	1–2.5 μg/kg	[105]
Maize	FB1	Direct competitive multi-channel immunoassay	Electrochemical	0.58 μg/L	[106]
Oat, wheat, rye, and maize	OTA, DON, FB1 and FB2	Competitive indirect immunoassay	Chemiluminescence	0.9–159 μg/kg	[107]
Wheat and maize	ZEN, T2 and FB1	Competitive assay format	Colorimetric	N/A	[108]
Wheat, Durum wheat, Barley, Maize, Oats	T-2 and HT-2 toxins	Competitive ELISA	Colorimetric	75 μg/kg	[109]
Maize, Rice, Hazelnut	AFB1	Non-competitive immunoassay	Fluorescence	70 pg/mL	[110]
Corn, Wheat, Feedstuff	ZEN, FB1, DON, AFB1	Suspension array immunoassay	Luminex 200 suspension array analyzer	0.51–6.0 ng/mL	[111]
Wheat and corn flours	DON, FB1 and OTA	Magnetic particle-based enzyme immunoassay	Colorimetric	0.1–5 ng/mL	[112]
Maize	FB1	Competitive immunoassay fluorescence correlation spectroscopy (FCS)	Fluorescence	1.0 mg/L	[113]
Maize and wheat	DON, ZEN, AFB1	QD@SiO_2-based immunoassay	Colorimetric	1.9–5.4 μg/kg	[114]
Wheat, Barley, Soybean, Rice, Maize, Rapeseed meal, Sunflower meal, Complete feeds	ZEN, DON, AFB1 and OTA	Enzyme-linked immunosorbent assay	Colorimetric	1.4–28 μg/kg	[115]
Wheat, Corn, Peanut, Feedstuff	AFB1, ZEN, DON, OTA, and FB1	Polyvinylidene fluoride (PVDF) membrane-based dot immunoassay	Densitometric analysis	20–1000 μg/kg	[116]
Corn, Wheat, Rice	ZEN	Indirect competitive phage ELISA anti-idiotypic VHH phage particles were applied to PD-IPCR	Colorimetric	6.5 pg/mL	[117]
Wheat and maize	DON and ZEN	Multiplex immunosorbent assay	Fluorescence	ZEN:100 μg/kg DON: 700 μg/kg	[118]
Wheat, Maize, Peanut Oil, Husked Rice	AFB1	Quantum dots and immunomagnetic beads	Atomic absorption spectroscopy (AAS)	0.04 μg/kg	[119]
Corn, wheat	ZEN	Competitive immunoassay integrated poly(dimethylsiloxane) (iPDMS)	Chemiluminescence	0.53 μg/kg	[120]
Wheat, Maize	a. DON, ZEN, AFB1, T2 and FB1 b. ZEN and AFB1	Fluorescent immunosorbent assay (FLISA) a. same plate (single-analyte multiplex, SAM) b. double-analyte multiplex (DAM)	Fluorescence	a. 0.4–10 μg/kg b. 1–1.8 μg/kg	[121]
Maize and cereal-based animal feeds	AFB1, ZEN, T-2 toxin	Multicolor-based immunochromatographic strip (ICS)	Optical	Visible detection limit: 0.5–30 ng/mL,	[122]
Maize, Wheat Rice	ZEN	Three kinds of lateral flow immunochromatographic assays (ICAs)	Colorimetric	6–60 μg/kg	[123]

A silver nanoparticle/carbon dot has been applied to develop a "turn on" pattern fluorescence quenching FLFIA (fluorescence lateral flow immunochromatographic assays) method for the qualitative and semi-quantitative detection of ZEN in maize, and wheat. This assay had a limit of detection (LOD) of 1–2.5 μg/kg for ZEN in cereal samples [105]. For ZEN detection, a monoclonal antibody (mAb) (2B10) was also prepared. The specific mAb showed no cross-reactivity with other groups of mycotoxins. A competitive microarray assay based on a novel solid supporting material, an integrated poly dimethylsiloxane (iPDMS), was proposed for qualitative and/or semiquantitative determination of ZEN providing a very low limit of quantification (LOQ) 1.02 μg/kg in cereal samples [120]. The variable domain of heavy-chain antibodies (VHHs) as alternative compounds to produce anti-idiotypic antibodies, which work as non-toxic surrogate reagents in immunoassay has also been applied. The proposed method proved to be reliable for the determination of ZEN in cereal samples with a LOD of 6.5 pg/mL. The use of antiidiotypic VHH phage as a non-toxic surrogate and the signal-amplification function of PCR makes it a promising method for actual ZEN analysis in corn, wheat, and rice [117]. In the same ingredients, three kinds of lateral flow immunochromatographic assays (ICAs) using colloidal gold, quantum dots, and polystyrene microspheres have also been used as labels for the detection of ZEN. The assays allow ZEN to be quantified within 20 min with LODs ranging between 6 and 60 μg/kg [123].

For sensitive detection of FB1 in maize, an immunoassay using single-molecule fluorescence correlation spectroscopy was developed. In comparison to conventional ELISA, this method showed high sensitivity, simplicity, a short analysis time, and low reagent and sample requirement. The LOD of this method for FB1 was 1 mg/L [113].

A single-step assay has been developed for the rapid detection of AFB1 in maize, rice, and hazelnut within 15 min with a LOD of 70 pg/mL. For this method, anti-immunocomplex (anti-IC) antibodies were used and the immunoassay was non-competitive showing the applicability of these parameters in the analysis of small molecule contaminants [110]. Furthermore, a chromatography-free method was found in the literature for the detection of AFB1 in cereals and oils through atomic absorption spectroscopy (AAS) using quantum dots and immunomagnetic beads. A magneto-controlled pre-treatment platform for automatic purification, labeling, and digestion was constructed and AFB1 detection through AAS was enabled using the proposed immunoassay which exhibits high sensitivity for AFB1 detection in wheat and maize, with a LOD of 0.04 μg/kg [119].

For the quantitative and simultaneous detection of different mycotoxins, various immunological methods have been assessed. Detection of ZEN, FB1, DON, and AFB1 in corn, and wheat, has been achieved by a suspension array. Suspension arrays have the advantages of sensitivity, rapidity, and accuracy. Signal responses are observed using red and green laser lights to achieve qualitative and quantitative detections. The LODs of the method were 0.51–6.0 ng/mL for the four mycotoxins [111]. Multiplex fluorescent immunosorbent assay (FLISA) using quantum dots (QDs)-based immunochemical techniques has been used for multi-contaminated cereal samples, allowing the simultaneous determination of all compounds. The mycotoxins DON, ZEN, AFB1, T-2, and FB1 were allocated to different wells of the same multi-well plate, and the sample was treated before being dispensed over the wells (single-analyte multiplex, SAM). Moreover, multi-contamination with ZEN and AFB1 was determined with the double-analyte multiplex (DAM). Two different specific antibodies were distributed in one single well and the mycotoxins ZEN and AFB1 were determined in wheat and maize, on the condition that their conjugates are labeled with QDs, which are fluorescent in different parts of the spectrum at two different wavelengths [121].

Another sensitive tool for the simultaneous quantitative determination of DON, ZEN, and AFB1 in cereal-based products in one single well of a microtiter plate, has been applied. This one is based on the use of a colloidal quantum dot enrobed into a silica shell (QD@SiO2) derivatives as a highly responsive label. Silica-coated quantum dots were prepared and subsequently modified via co-hydrolysis with tetraethylorthosilicate (TEOS) and various organosilane reagents. The LODs were 1.9–5.4 μg/kg for the three mycotoxins [114]. Moreover, another study proposed the development of a polyvinylidene fluoride (PVDF) membrane-based dot immunoassay for the rapid and simultaneous detection of AFB1, ZEN, DON, OTA, and FB1 in corn, and wheat. The LODs of the method for mycotoxins were 20–1000 μg/kg [116].

Cereal contamination with ZEN and DON, was identified using Cd-based QDs as labels, while an imprinted BSA was immobilized on a microwell plate. This technique involved putting silica on green- and red-emitting QDs to turn them hydrophilic, before coupling with mycotoxin-protein occurs. The ZEN detection cut-off level varies depending on cereal origin. On the other hand, the cut-off level for DON is considerably lower when compared to its permissible limits [118]. The multi-mycotoxins (AFB1, ZEN, and T-2 toxin) determination in cereals has also been accomplished using a multicolor immunochromatographic strip (ICS). On this method, three monoclonal antibodies are bound to three different colored nanoparticles to act as immunoassay probes and the three mycotoxins may be quantified at the same time according to color decrease [122].

3.3. Biosensors

Biosensors consist of various elements such as a molecularly imprinted polymer (MIP), an aptamer, a DNA/RNA molecule, an enzyme, a tissue, living cells, and antibodies. A transducer is also necessary to connect these parts, which transforms the observed physical or chemical changes into a quantifiable signal. Depending on the signal transduction mechanism and the applied recognition elements, three categories of biosensors exist: optical, electrochemical, and piezoelectric. Immunosensors are of the most commonly used analytical methods for mycotoxin detection, although other cutting-edge methods such as MIP-based sensors are available. Antibodies, antigens, and their fragments, are used for biomolecular recognition in immunosensors. The essential premise behind all immunosensors is that the precise binding of the immobilized components in the sample results in the production of an analytical signal that is affected by the concentration of the target analyte. Labeled and label-free immunosensors combined with different transducers have been considerably developed for mycotoxin assessment [124–126]. The various mycotoxin detection sensors that have been created over the previous ten years are presented in Table 3.

Table 3. Biosensor methods used in the analysis of mycotoxins.

Type of Cereal	Mycotoxins	Method	Detection Method	LOD	Ref.
Maize, Rice	ZEN	Direct binding surface of MSNs-NH2 and the aptamer-FAM (molecular recognition probe)	Fluorescence	0.012 ng/mL	[127]
Maize and cereals feedstuff	ZEN	Flow-through fluorescence sensor	Fluorescence	15 µg/kg	[128]
Wheat and maize samples	DON	Competitive immunoassay	Optical immunosensor White Light Reflectance Spectroscopy (WLRS)	62.5 µg/kg	[129]
Corn	ZEN	SERS-based test strip bimetallic core–shell Au@AgNPs with embedded reporter molecules (4-MBA) as the SERS nanoprobe	Raman spectrometry	3.6 µg/kg	[130]
Corn	OTA	Differential pulse voltammetric aptasensor based on hybridization chain reaction	Electrochemical	2 pg·/mL	[131]
Corn, Wheat	DON	Indirect competitive immunoassay s	Fluorescence	0.16 µg/L	[132]

For ZEN detection in maize and rice, a fluorometric assay based on mesoporous silica nanoparticles (MSNs-NH2) as a positive charge reactor and an aptamer-FAM (6-carboxy-fluorescein-labeled aptamer) as a signal probe (capture probe and negative charge reactor), was tested respectively. The proposed assay had high recognition specificity, low LOD (0.012 ng/mL), and a wide linear range (0.005–150 ng/mL) [127]. Furthermore, a multi-commutated flow-through optosensor in different cereal samples was developed to quantify ZEN. The mycotoxin was retained and pre-concentrated on C18 silica gel, and the use of the multi-commutated flow manifold allowed the automated retention/desorption of ZEN on the solid microbeads using appropriate carrier/eluting solutions. The native fluorescence of ZEN was recorded on the solid phase at λexc/λem of 265/465 nm/nm. A QuEChERS procedure was used as a clean-up step of ZEN from different cereal samples. Recovery studies were performed to assess the accuracy of the method, obtaining recovery yields between 93% and 107% in all the analyzed samples (maize and cereals feedstuff) and the LOD was 15 µg/kg [128]. Furthermore, a surface-enhanced Raman scattering (SERS)-based test strip was proposed for the detection of ZEN, showing simplicity, rapidity, and high sensitivity. Core-shell Au@AgNPs with embedded reporter molecules (4-MBA) was synthesized as SERS nanoprobe, which exhibited excellent SERS signals and high stability. The detection range of ZEN for corn samples was 10–1000 µg/kg while the LOD of the method was 3.6 µg/kg [130].

DON detection in crop samples (wheat and maize) has been achieved using a white light reflectance spectroscopy (WLRS) optical immunosensor. It was proved to be a fast and high-sensitivity assay for the assessment of contamination in the whole grain [129].

For the determination of the mycotoxin OTA, a differential pulse voltammetric aptasensor based on hybridization chain reaction (HCR) was developed. The assay was successfully applied to the determination of OTA in cereal samples with a detection limit of 2 pg/mL [131]. A portable and reusable optofluidic immunosensor OIP-v2 was developed for rapid and sensitive on-site detection of DON using DON-BSA modified bio-probes as biorecognition elements. The OIP-v2 was used for the detection of DON with high sensitivity, accuracy, and rapidity. The LOD of DON was 0.16 μg/L [132].

3.4. Spectroscopic Methods FT-NIR

Infrared (IR) spectroscopy-based methods are the most promising for the detection of mycotoxins since they require small samples and limited technical expertise. Moreover, such techniques are cheap and need no sample pre-treatment. Identification of mycotoxin contamination in crops is commonly carried out using spectroscopic techniques [133]. Mid-infrared (MIR) spectroscopy is specified for molecular vibrations while standard NIR spectroscopy determines the molecular overtones and combined vibrations of chemical bonds. All spectra produced by overtones and mixed vibrations seen in the NIR range are challenging to decipher for specific constituents present in a sample [134]. Chemometrics can be used for direct information extraction from the data, which solves the upcoming necessity of mathematical processing to extract chemicals and linked information in the assessment of NIR and MIR spectra. Three phases are often involved in NIR or MIR spectroscopy using chemometrics: spectral pre-processing, multivariate model construction for calibration, and model transfer [135]. Qualitative and quantitative methodologies are used for NIR and MIR spectroscopic model development. Some examples of qualitative techniques include principal component analysis (PCA), cluster analysis (CA), and linear discriminant analysis (LDA). On the other hand, principal component regression (PCR), multiple linear regression, and partial least squares (PLS) are common methodologies for quantitative multivariate calibration (MLR).

FB1 and FB2 concentrations in maize meal were first analyzed for mycotoxins using FT-IR as a quick way to distinguish contaminated meals [136]. Based on an optimized feature model for NIR spectroscopy, a quantitative assay for AFB1 in maize has been suggested. The potential of NIR spectroscopy in conjunction with chemometric techniques for the quick and accurate quantitative detection of the AFB1 in maize was demonstrated using a portable NIR spectroscopy device to evaluate maize samples with varied degrees of contamination. To effectively mine the wavelengths of the NIR spectra, different variable selection algorithms were used. After the screening, the wavelength variables were utilized to create a support vector machine (SVM) and a partial least squares (PLS) test model, respectively, to measure AFB1 in maize. As a result, by using a nonlinear SVM detection model, the characteristics of NIR spectra are beneficial for the rapid and accurate testing of the AFB1 in maize [136].

NIR spectroscopy is used for rapid ZEN identification in wheat grains. First, using Savitzky–Golay smoothing (SG-smoothing) and multiple scattering correction (MSC), the collected original NIR spectra were denoised, smoothed, and scatter-corrected, before normalized. Random frog (RF), successive projections algorithm (SPA), least absolute shrinkage, and selection operator were the three algorithms utilized to choose variables from the pre-processed NIR spectra (LASSO). In order to achieve the quantitative detection of the ZEN in wheat grains, SVM models were built based on the feature variables extracted by the aforementioned techniques and the LASSO-SVM model's prediction effect proved to be more accurate [137].

Total FBs (FB1 + FB2) and ZEN in Brazilian maize have also been measured using NIR [138]. There were three regression models used: one for FB1 with 18 principal components (PCs), one for FB2 with 10 PCs, and one for ZEN with 7 PCs. As internal validation, a partial least squares regression technique with full cross-validation was used. When FBs and ZEN were assessed using various assessing calibrations NIR values did not differ significantly compared to reference values LC-MS/MS values, presenting NIR as a reliable method for quick detection of FBs and ZEN in corn [138].

Finally, NIR and FT-NIR were used to evaluate their applicability and efficiency for the analysis of Brazilian wheat flour samples contaminated with DON, with partial least-squares discriminant analysis (PLS-DA) and principal component analysis-linear discriminant analysis (PC-LDA) used as discriminatory methods. Validation samples through PLS-DA showed correct classification rates in the range of 85–87.5% with an error of 10–15% error. For PC-LDA, the hit rate was over 85% with an error of 10–15% demonstrating that NIR is an excellent alternative method for the classification of wheat flour samples according to DON content [139].

4. Conclusions and Outlook

The presence of mycotoxins in agricultural products of animal feeds animals poses a health risk to both livestock and consumers. Due to their toxic nature, their prompt detection of mycotoxins is critical to the food industry including aquaculture. During the last decade, analytical methods used for mycotoxin detection include the use of chromatographic, immunological, and spectroscopic (NIR) methods as well as biosensors. Chromatographic methods using LC-MS/MS are more sensitive but require technical competence and higher time investment. Immunological methods such as ELISA and biosensors are less sensitive and reliable but are simpler to use by non-specialized personnel directly in the field and without the requirement of laboratory infrastructure. The FT-NIR spectroscopic method, which has been extensively utilized in recent years for mycotoxin detection, is relatively simple and eco-friendly; however, it requires expensive infrastructure and complex chemometrics and mathematical calculations for its development. As mycotoxins will probably continue to be a problem for the aquaculture industry, to ensure the safety of aquafeeds and produced healthy seafood, it becomes necessary to develop certain detection methods, especially those that can be used on-site.

Funding: This research is co-funded by Greece and the European Union under the Fisheries and Maritime Operational Program 2014–2020 (75% EMFF contribution, 25% National Contribution).

Data Availability Statement: Not applicable.

Conflicts of Interest: The authors declare no conflict of interest.

References

1. FAO. *The State of World Fisheries and Aquaculture 2022. Towards Blue Transformation*; FAO: Rome, Italy, 2022.
2. Edwards, P.; Zhang, W.; Belton, B.; Little, D.C. Misunderstandings, myths and mantras in aquaculture: Its contribution to world food supplies has been systematically over reported. *Mar. Policy* **2019**, *106*, 103547. [CrossRef]
3. Ottinger, M.; Clauss, K.; Kuenzer, C. Aquaculture: Relevance, distribution, impacts and spatial assessments–A review. *Ocean Coast. Manag.* **2016**, *119*, 244–266. [CrossRef]
4. Springmann, M.; Clark, M.; Mason-D'Croz, D.; Wiebe, K.; Bodirsky, B.L.; Lassaletta, L.; De Vries, W.; Vermeulen, S.J.; Herrero, M.; Carlson, K.M. Options for keeping the food system within environmental limits. *Nature* **2018**, *562*, 519–525. [CrossRef]
5. Willett, W.; Rockström, J.; Loken, B.; Springmann, M.; Lang, T.; Vermeulen, S.; Garnett, T.; Tilman, D.; DeClerck, F.; Wood, A. Food in the Anthropocene: The EAT–Lancet Commission on healthy diets from sustainable food systems. *Lancet* **2019**, *393*, 447–492. [CrossRef]
6. Newton, R.; Maiolo, S.; Malcorps, W.; Little, D. Life cycle inventories of marine ingredients. *Aquaculture* **2023**, *565*, 739096. [CrossRef]
7. Miles, R.D.; Chapman, F.A. The benefits of fish meal in aquaculture diets. FA122/FA122, 5/2006. *EDIS* **2006**, *12*, 1–7.
8. Naylor, R.L.; Hardy, R.W.; Buschmann, A.H.; Bush, S.R.; Cao, L.; Klinger, D.H.; Little, D.C.; Lubchenco, J.; Shumway, S.E.; Troell, M. A 20-year retrospective review of global aquaculture. *Nature* **2021**, *591*, 551–563. [CrossRef] [PubMed]
9. IndexMundi—Country Facts. Available online: https://www.indexmundi.com/agriculture/?commodity=fish-meal&graph=production (accessed on 23 December 2022).
10. Maiolo, S.; Parisi, G.; Biondi, N.; Lunelli, F.; Tibaldi, E.; Pastres, R. Fishmeal partial substitution within aquafeed formulations: Life cycle assessment of four alternative protein sources. *Int. J. Life Cycle Assess.* **2020**, *25*, 1455–1471. [CrossRef]

11. Elumalai, P.; Kurian, A.; Lakshmi, S.; Faggio, C.; Esteban, M.A.; Ringø, E. Herbal immunomodulators in aquaculture. *Rev. Fish. Sci. Aquac.* **2020**, *29*, 33–57. [CrossRef]
12. Gatlin, D.M., III; Barrows, F.T.; Brown, P.; Dabrowski, K.; Gaylord, T.G.; Hardy, R.W.; Herman, E.; Hu, G.; Krogdahl, Å.; Nelson, R. Expanding the utilization of sustainable plant products in aquafeeds: A review. *Aquac. Res.* **2007**, *38*, 551–579. [CrossRef]
13. Hardy, R.W. Utilization of plant proteins in fish diets: Effects of global demand and supplies of fishmeal. *Aquac. Res.* **2010**, *41*, 770–776. [CrossRef]
14. Sales, J. The effect of fish meal replacement by soyabean products on fish growth: A meta-analysis. *Br. J. Nutr.* **2009**, *102*, 1709–1722. [CrossRef]
15. Hixson, S.M. Fish nutrition and current issues in aquaculture: The balance in providing safe and nutritious seafood, in an environmentally sustainable manner. *J. Aquac. Res. Dev.* **2014**, *5*, 1–10. [CrossRef]
16. Samuelsen, T.A.; Hillestad, M.; Jacobsen, H.J.; Hjertnes, T.J.; Sixten, H.J. Physical feed quality and starch content causes a biological response in Atlantic salmon (*Salmo salar* L.). *Aquac. Rep.* **2021**, *21*, 100791. [CrossRef]
17. Troell, M.; Naylor, R.L.; Metian, M.; Beveridge, M.; Tyedmers, P.H.; Folke, C.; Arrow, K.J.; Barrett, S.; Crépin, A.-S.; Ehrlich, P.R. Does aquaculture add resilience to the global food system? *Proc. Natl. Acad. Sci. USA* **2014**, *111*, 13257–13263. [CrossRef]
18. Sicuro, B.; Gai, F.; Daprà, F.; Palmegiano, G.B. Hybrid sturgeon 'AL'(Acipenser naccarii× Acipenser baeri) diets: The use of alternative plant protein sources. *Aquac. Res.* **2012**, *43*, 161–166. [CrossRef]
19. Bonaldo, A.; Parma, L.; Mandrioli, L.; Sirri, R.; Fontanillas, R.; Badiani, A.; Gatta, P.P. Increasing dietary plant proteins affects growth performance and ammonia excretion but not digestibility and gut histology in turbot (*Psetta maxima*) juveniles. *Aquaculture* **2011**, *318*, 101–108. [CrossRef]
20. Nogales-Mérida, S.; Tomás-Vidal, A.; Moñino-López, A.; Jover-Cerdá, M.; Martínez-Llorens, S. Pea protein concentrate in diets for sharpsnout sea bream (*Diplodus puntazzo*): Effects on growth and health status. *Arch. Anim. Nutr.* **2016**, *70*, 488–502. [CrossRef] [PubMed]
21. Parma, L.; Candela, M.; Soverini, M.; Turroni, S.; Consolandi, C.; Brigidi, P.; Mandrioli, L.; Sirri, R.; Fontanillas, R.; Gatta, P.P. Next-generation sequencing characterization of the gut bacterial community of gilthead sea bream (*Sparus aurata*, L.) fed low fishmeal based diets with increasing soybean meal levels. *Anim. Feed Sci. Technol.* **2016**, *222*, 204–216. [CrossRef]
22. Krogdahl, Å.; Penn, M.; Thorsen, J.; Refstie, S.; Bakke, A.M. Important antinutrients in plant feedstuffs for aquaculture: An update on recent findings regarding responses in salmonids. *Aquac. Res.* **2010**, *41*, 333–344. [CrossRef]
23. Oliva-Teles, A. Nutrition and health of aquaculture fish. *J. Fish Dis.* **2012**, *35*, 83–108. [CrossRef]
24. Oliva-Teles, A.; Enes, P.; Peres, H. Replacing fishmeal and fish oil in industrial aquafeeds for carnivorous fish. In *Feed and Feeding Practices in Aquaculture*; Woodhead Publishing Series in Food Science, Technology and Nutrition; Woodhead Publishing: Sawston, CA, USA, 2015; pp. 203–233.
25. Kokou, F.; Fountoulaki, E. Aquaculture waste production associated with antinutrient presence in common fish feed plant ingredients. *Aquaculture* **2018**, *495*, 295–310. [CrossRef]
26. Gonçalves, R.A.; Schatzmayr, D.; Albalat, A.; Mackenzie, S. Mycotoxins in aquaculture: Feed and food. *Rev. Aquac.* **2020**, *12*, 145–175. [CrossRef]
27. Haque, M.A.; Wang, Y.; Shen, Z.; Li, X.; Saleemi, M.K.; He, C. Mycotoxin contamination and control strategy in human, domestic animal and poultry: A review. *Microb. Pathog.* **2020**, *142*, 104095. [CrossRef]
28. Palumbo, R.; Crisci, A.; Venâncio, A.; Cortiñas Abrahantes, J.; Dorne, J.-L.; Battilani, P.; Toscano, P. Occurrence and co-occurrence of mycotoxins in cereal-based feed and food. *Microorganisms* **2020**, *8*, 74. [CrossRef]
29. Streit, E.; Naehrer, K.; Rodrigues, I.; Schatzmayr, G. Mycotoxin occurrence in feed and feed raw materials worldwide: Long-term analysis with special focus on Europe and Asia. *J. Sci. Food Agric.* **2013**, *93*, 2892–2899. [CrossRef]
30. International Agency for Research on Cancer. *A Review of Human Carcinogens. F. Chemical Agents and Related Occupations: IARC Monographs on the Evaluation of Carcinogenic Risks to Humans*; International Agency for Research on Cancer: Lyon, France, 2012; p. 100F.
31. World Health Organization; International Agency for Research on Cancer. Some naturally occurring substances: Food items and constituents, heterocyclic aromatic amines and mycotoxins. In *IARC Monographs on the Evaluation of the Carcinogenic Risk of Chemicals to Humans*; World Health Organization: Geneva, Switzerland, 1993; Volume 56.
32. Tomatis, L.; Wilbourn, J. Evaluation of carcinogenic risk to humans: The experience of IARC. In *New Frontiers in Cancer Causation*; CRC Press: Boca Raton, FL, USA, 1993; pp. 371–387.
33. Alassane-Kpembi, I.; Schatzmayr, G.; Taranu, I.; Marin, D.; Puel, O.; Oswald, I.P. Mycotoxins co-contamination: Methodological aspects and biological relevance of combined toxicity studies. *Crit. Rev. Food Sci. Nutr.* **2017**, *57*, 3489–3507. [CrossRef]
34. Lee, H.J.; Ryu, D. Worldwide occurrence of mycotoxins in cereals and cereal-derived food products: Public health perspectives of their co-occurrence. *J. Agric. Food Chem.* **2017**, *65*, 7034–7051. [CrossRef] [PubMed]
35. Santos Pereira, C.; Cunha, S.C.; Fernandes, J.O. Prevalent mycotoxins in animal feed: Occurrence and analytical methods. *Toxins* **2019**, *11*, 290. [CrossRef] [PubMed]
36. Marijani, E.; Kigadye, E.; Okoth, S. Occurrence of fungi and mycotoxins in fish feeds and their impact on fish health. *Int. J. Microbiol.* **2019**, *2019*, 1–17. [CrossRef]
37. Oliveira, M.; Vasconcelos, V. Occurrence of mycotoxins in fish feed and its effects: A review. *Toxins* **2020**, *12*, 160. [CrossRef]
38. Zain, M.E. Impact of mycotoxins on humans and animals. *J. Saudi Chem. Soc.* **2011**, *15*, 129–144. [CrossRef]

39. Koletsi, P.; Schrama, J.W.; Graat, E.A.M.; Wiegertjes, G.F.; Lyons, P.; Pietsch, C. The Occurrence of Mycotoxins in Raw Materials and Fish Feeds in Europe and the Potential Effects of Deoxynivalenol (DON) on the Health and Growth of Farmed Fish Species—A Review. *Toxins* **2021**, *13*, 403. [CrossRef]
40. Da Rocha, M.E.B.; Freire, F.d.C.O.; Maia, F.E.F.; Guedes, M.I.F.; Rondina, D. Mycotoxins and their effects on human and animal health. *Food Control* **2014**, *36*, 159–165. [CrossRef]
41. Directive 2002/32/EC of the European Parliament and of the Council of 7 May 2002 on Undesirable Substances in Animal Feed. Available online: https://webarchive.nationalarchives.gov.uk/eu-exit/https://eur-lex.europa.eu/legal-content/EN/TXT/?uri=CELEX:02002L0032-20191128 (accessed on 28 December 2022).
42. Alexander, J.; Benford, D.; Boobis, A.; Ceccatelli, S.; Cottrill, B.; Cravedi, J.-P.; Di Domenico, A.; Doerge, D.; Dogliotti, E.; Edler, L.; et al. Scientific Opinion on Ergot alkaloids in food and feed. *EFSA J.* **2012**, *10*, 2798. [CrossRef]
43. Commission recommendation on the presence of deoxynivalenol, zearalenone, ochratoxin A, T-2 and HT-2 and fumonisins in products intended for animal feeding (Text with EEA relevance) At the request of the Commission the European Food Safety Authority (EFSA) adopted opinions on the myco-toxins deoxynivalenol. *Off. J. Eur. Union* **2006**, *229*, 7–9.
44. EFSA Panel on Contaminants in the Food Chain (CONTAM). Scientific Opinion on the risks for animal and public health related to the presence of T-2 and HT-2 toxin in food and feed. *EFSA J.* **2011**, *9*, 2481. [CrossRef]
45. Singh, J.; Mehta, A. Rapid and sensitive detection of mycotoxins by advanced and emerging analytical methods: A review. *Food Sci. Nutr.* **2020**, *8*, 2183–2204. [CrossRef]
46. Janik, E.; Niemcewicz, M.; Podogrocki, M.; Ceremuga, M.; Gorniak, L.; Stela, M.; Bijak, M. The existing methods and novel approaches in mycotoxins' detection. *Molecules* **2021**, *26*, 3981. [CrossRef]
47. Wagner, C. Critical practicalities in sampling for mycotoxins in feed. *J. AOAC Int.* **2015**, *98*, 301–308. [CrossRef] [PubMed]
48. Xie, L.; Chen, M.; Ying, Y. Development of methods for determination of aflatoxins. *Crit. Rev. Food Sci. Nutr.* **2016**, *56*, 2642–2664. [CrossRef]
49. The European Commission Commission Regulation (EC) No 691/2013 of 19 July 2013 amending Regulation (EC) No 152/2009 as regards methods of sampling and analysis. *Off. J. Eur. Union* **2013**, *197*, 1–12.
50. Turner, N.W.; Bramhmbhatt, H.; Szabo-Vezse, M.; Poma, A.; Coker, R.; Piletsky, S.A. Analytical methods for determination of mycotoxins: An update (2009–2014). *Anal. Chim. Acta* **2015**, *901*, 12–33. [CrossRef]
51. European Commision. Commission Regulation (EC) No 401/2006 of 23 February 2006 laying down the methods of sampling and analysis for the official control of the levels of mycotoxins in foodstuffs. *Off. J. Eur. Union* **2006**, *70*, 12–34.
52. Maestroni, B.; Cannavan, A. Sampling strategies to control mycotoxins. In *Determining Mycotoxins and Mycotoxigenic Fungi in Food and Feed*; Woodhead Publishing: Sawston, CA, USA, 2011; pp. 3–36.
53. Turner, N.W.; Subrahmanyam, S.; Piletsky, S.A. Analytical methods for determination of mycotoxins: A review. *Anal. Chim. Acta* **2009**, *632*, 168–180. [CrossRef]
54. Whitaker, T.B. Standardisation of mycotoxin sampling procedures: An urgent necessity. *Food Control* **2003**, *14*, 233–237. [CrossRef]
55. Miraglia, M.; De Santis, B.; Minardi, V.; Debegnach, F.; Brera, C. The role of sampling in mycotoxin contamination: An holistic view. *Food Addit. Contam.* **2005**, *22*, 31–36. [CrossRef] [PubMed]
56. Krska, R.; Welzig, E.; Berthiller, F.; Molinelli, A.; Mizaikoff, B. Advances in the analysis of mycotoxins and its quality assurance. *Food Addit. Contam.* **2005**, *22*, 345–353. [CrossRef]
57. Razzazi-Fazeli, E.; Reiter, E.V. Sample preparation and clean up in mycotoxin analysis: Principles, applications and recent developments. In *Determining Mycotoxins and Mycotoxigenic Fungi in Food and Feed*; Woodhead Publishing: Sawston, CA, USA, 2011; pp. 37–70.
58. Kafouris, D.; Christofidou, M.; Christodoulou, M.; Christou, E.; Ioannou-Kakouri, E. A validated UPLC-MS/MS multi-mycotoxin method for nuts and cereals: Results of the official control in Cyprus within the EU requirements. *Food Agric. Immunol.* **2017**, *28*, 90–108. [CrossRef]
59. Kovač, M.; Nevistić, A.; Kovač, T.; Babić, J.; Šarić, A.; Miličević, B.; Panjičko, M.; Šarkanj, B. Development and Validation of an UHPLC-MS/MS Method for the Simultaneous Determination of 11 EU-Regulated Mycotoxins in Selected Cereals. *J. Fungi* **2022**, *8*, 665. [CrossRef] [PubMed]
60. Nathanail, A.V.; Syvähuoko, J.; Malachová, A.; Jestoi, M.; Varga, E.; Michlmayr, H.; Adam, G.; Sieviläinen, E.; Berthiller, F.; Peltonen, K. Simultaneous determination of major type A and B trichothecenes, zearalenone and certain modified metabolites in Finnish cereal grains with a novel liquid chromatography-tandem mass spectrometric method. *Anal. Bioanal. Chem.* **2015**, *407*, 4745–4755. [CrossRef]
61. Silva, A.S.; Brites, C.; Pouca, A.V.; Barbosa, J.; Freitas, A. UHPLC-ToF-MS method for determination of multi-mycotoxins in maize: Development and validation. *Curr. Res. Food Sci.* **2019**, *1*, 1–7. [CrossRef] [PubMed]
62. Cortese, M.; Gigliobianco, M.R.; Magnoni, F.; Censi, R.; Di Martino, P. Compensate for or minimize matrix effects? Strategies for overcoming matrix effects in liquid chromatography-mass spectrometry technique: A tutorial review. *Molecules* **2020**, *25*, 3047. [CrossRef]
63. Furey, A.; Moriarty, M.; Bane, V.; Kinsella, B.; Lehane, M. Ion suppression; a critical review on causes, evaluation, prevention and applications. *Talanta* **2013**, *115*, 104–122. [CrossRef] [PubMed]
64. Varzakas, T.; Tzia, C. *Food Engineering Handbook: Food Process Engineering*; CRC Press: Boca Raton, FL, USA, 2014; ISBN 1482261669.

65. Prado, J.; Rostagno, M. *Natural Product Extraction: Principles and Applications*; Royal Society of Chemistry: London, UK, 2022; ISBN 1839165901.
66. Annunziata, L.; Stramenga, A.; Visciano, P.; Schirone, M.; De Colli, L.; Colagrande, M.N.; Campana, G.; Scortichini, G. Simultaneous determination of aflatoxins, T-2 and HT-2 toxins, and fumonisins in cereal-derived products by QuEChERS extraction coupled with LC-MS/MS. *Anal. Bioanal. Chem.* **2017**, *409*, 5143–5155. [CrossRef]
67. Bolechová, M.; Benešová, K.; Běláková, S.; Čáslavský, J.; Pospíchalová, M.; Mikulíková, R. Determination of seventeen mycotoxins in barley and malt in the Czech Republic. *Food Control* **2015**, *47*, 108–113. [CrossRef]
68. González-Jartín, J.M.; Alfonso, A.; Sainz, M.J.; Vieytes, M.R.; Botana, L.M. Multi-detection method for mycotoxins with a modified QuEChERS extraction in feed and development of a simple detoxification procedure. *Anim. Feed Sci. Technol.* **2021**, *272*, 114745. [CrossRef]
69. Kim, H.; Baek, E.J.; Shin, B.G.; Kim, H.J.; Kim, J.-E. In-house validation of an efficient and rapid procedure for the simultaneous determination and monitoring of 23 mycotoxins in grains in Korea. *Toxins* **2022**, *14*, 457. [CrossRef]
70. Rausch, A.-K.; Brockmeyer, R.; Schwerdtle, T. Development and validation of a QuEChERS-based liquid chromatography tandem mass spectrometry multi-method for the determination of 38 native and modified mycotoxins in cereals. *J. Agric. Food Chem.* **2020**, *68*, 4657–4669. [CrossRef]
71. Tamura, M.; Mochizuki, N.; Nagatomi, Y.; Harayama, K.; Toriba, A.; Hayakawa, K. A method for simultaneous determination of 20 Fusarium toxins in cereals by high-resolution liquid chromatography-orbitrap mass spectrometry with a pentafluorophenyl column. *Toxins* **2015**, *7*, 1664–1682. [CrossRef]
72. Zhou, Q.; Li, F.; Chen, L.; Jiang, D. Quantitative analysis of 10 mycotoxins in wheat flour by ultrahigh performance liquid chromatography-tandem mass spectrometry with a modified QuEChERS strategy. *J. Food Sci.* **2016**, *81*, T2886–T2890. [CrossRef] [PubMed]
73. Scott, P.M.; Trucksess, M.W. Application of immunoaffinity columns to mycotoxin analysis. *J. AOAC Int.* **1997**, *80*, 941–950. [CrossRef]
74. Gab-Allah, M.A.; Tahoun, I.F.; Yamani, R.N.; Rend, E.A.; Shehata, A.B. Eco-friendly and sensitive analytical method for determination of T-2 toxin and HT-2 toxin in cereal products using UPLC-MS/MS. *J. Food Compos. Anal.* **2022**, *107*, 104395. [CrossRef]
75. Göbel, R.; Lusky, K. Simultaneous determination of aflatoxins, ochratoxin A, and zearalenone in grains by new immunoaffinity column/liquid chromatography. *J. AOAC Int.* **2004**, *87*, 411–416. [CrossRef] [PubMed]
76. Pascale, M.; Panzarini, G.; Visconti, A. Determination of HT-2 and T-2 toxins in oats and wheat by ultra-performance liquid chromatography with photodiode array detection. *Talanta* **2012**, *89*, 231–236. [CrossRef]
77. De Boevre, M.; Di Mavungu, J.D.; Maene, P.; Audenaert, K.; Deforce, D.; Haesaert, G.; Eeckhout, M.; Callebaut, A.; Berthiller, F.; Van Peteghem, C. Development and validation of an LC-MS/MS method for the simultaneous determination of deoxynivalenol, zearalenone, T-2-toxin and some masked metabolites in different cereals and cereal-derived food. *Food Addit. Contam. Part A* **2012**, *29*, 819–835. [CrossRef]
78. Boulard, L.; Parrhysius, P.; Jacobs, B.; Dierkes, G.; Wick, A.; Buchmeier, G.; Koschorreck, J.; Ternes, T.A. Development of an analytical method to quantify pharmaceuticals in fish tissues by liquid chromatography-tandem mass spectrometry detection and application to environmental samples. *J. Chromatogr. A* **2020**, *1633*, 461612. [CrossRef]
79. Rios, A.; Zougagh, M.; Bouri, M. Magnetic (nano) materials as an useful tool for sample preparation in analytical methods. A review. *Anal. Methods* **2013**, *5*, 4558–4573. [CrossRef]
80. Mashhadizadeh, M.H.; Amoli-Diva, M.; Pourghazi, K. Magnetic nanoparticles solid phase extraction for determination of ochratoxin A in cereals using high-performance liquid chromatography with fluorescence detection. *J. Chromatogr. A* **2013**, *1320*, 17–26. [CrossRef]
81. La Barbera, G.; Capriotti, A.L.; Cavaliere, C.; Foglia, P.; Montone, C.M.; Chiozzi, R.Z.; Laganà, A. A Rapid Magnetic Solid Phase Extraction Method Followed by Liquid Chromatography-Tandem Mass Spectrometry Analysis for the Determination of Mycotoxins in Cereals. *Toxins* **2017**, *9*, 147. [CrossRef]
82. Lucci, P.; Derrien, D.; Alix, F.; Pérollier, C.; Bayoudh, S. Molecularly imprinted polymer solid-phase extraction for detection of zearalenone in cereal sample extracts. *Anal. Chim. Acta* **2010**, *672*, 15–19. [CrossRef] [PubMed]
83. Cavaliere, C.; Antonelli, M.; Cerrato, A.; La Barbera, G.; Laganà, A.; Laus, M.; Piovesana, S.; Capriotti, A.L. A novel magnetic molecular imprinted polymer for selective extraction of zearalenone from cereal flours before liquid chromatography-tandem mass spectrometry determination. *Toxins* **2019**, *11*, 493. [CrossRef] [PubMed]
84. Xie, Z.; Lu, G.; Yan, Z.; Liu, J.; Wang, P.; Wang, Y. Bioaccumulation and trophic transfer of pharmaceuticals in food webs from a large freshwater lake. *Environ. Pollut.* **2017**, *222*, 356–366. [CrossRef]
85. Krska, R.; Stubbings, G.; Macarthur, R.; Crews, C. Simultaneous determination of six major ergot alkaloids and their epimers in cereals and foodstuffs by LC–MS–MS. *Anal. Bioanal. Chem.* **2008**, *391*, 563–576. [CrossRef]
86. Soleimany, F.; Jinap, S.; Faridah, A.; Khatib, A. A UPLC–MS/MS for simultaneous determination of aflatoxins, ochratoxin A, zearalenone, DON, fumonisins, T-2 toxin and HT-2 toxin, in cereals. *Food Control* **2012**, *25*, 647–653. [CrossRef]
87. Agriopoulou, S.; Stamatelopoulou, E.; Varzakas, T. Advances in occurrence, importance, and mycotoxin control strategies: Prevention and detoxification in foods. *Foods* **2020**, *9*, 137. [CrossRef]

88. Bi, S.; Xu, J.; Yang, X.; Zhang, P.; Lian, K.; Ma, L. An HPLC-MS/MS Method Using a Multitoxin Clean up Column for Analysis of Seven Mycotoxins in Aquafeeds. *J. AOAC Int.* **2022**, *105*, 107–114. [CrossRef]
89. Moreno-González, D.; Huertas-Pérez, J.F.; García-Campaña, A.M.; Gámiz-Gracia, L. Determination of carbamates in edible vegetable oils by ultra-high performance liquid chromatography–tandem mass spectrometry using a new clean-up based on zirconia for QuEChERS methodology. *Talanta* **2014**, *128*, 299–304. [CrossRef]
90. Kim, D.-B.; Jung, Y.S.; Nam, T.G.; Lee, S.; Yoo, M. Simultaneous determination of trichothecene mycotoxins in cereals by LC-MS/MS. *Food Sci. Biotechnol.* **2022**, *31*, 165–174. [CrossRef]
91. Lijalem, Y.G.; Gab-Allah, M.A.; Choi, K.; Kim, B. Development of isotope dilution-liquid chromatography/tandem mass spectrometry for the accurate determination of zearalenone and its metabolites in corn. *Food Chem.* **2022**, *384*, 132483. [CrossRef]
92. Gab-Allah, M.A.; Choi, K.; Kim, B. Development of isotope dilution-liquid chromatography/tandem mass spectrometry for the accurate determination of type-A trichothecenes in grains. *Food Chem.* **2021**, *344*, 128698. [CrossRef] [PubMed]
93. Meerpoel, C.; Vidal, A.; di Mavungu, J.D.; Huybrechts, B.; Tangni, E.K.; Devreese, M.; Croubels, S.; De Saeger, S. Development and validation of an LC–MS/MS method for the simultaneous determination of citrinin and ochratoxin a in a variety of feed and foodstuffs. *J. Chromatogr. A* **2018**, *1580*, 100–109. [CrossRef] [PubMed]
94. Liu, Z.; Wu, H.-L.; Xie, L.-X.; Hu, Y.; Fang, H.; Sun, X.-D.; Wang, T.; Xiao, R.; Yu, R.-Q. Chemometrics-enhanced liquid chromatography-full scan-mass spectrometry for interference-free analysis of multi-class mycotoxins in complex cereal samples. *Chemom. Intell. Lab. Syst.* **2017**, *160*, 125–138. [CrossRef]
95. Skendi, A.; Irakli, M.N.; Papageorgiou, M.D. Optimized and validated high-performance liquid chromatography method for the determination of deoxynivalenol and aflatoxins in cereals. *J. Sep. Sci.* **2016**, *39*, 1425–1432. [CrossRef] [PubMed]
96. Vidal, J.C.; Duato, P.; Bonel, L.; Castillo, J.R. Molecularly imprinted on-line solid-phase extraction coupled with fluorescence detection for the determination of ochratoxin A in wheat samples. *Anal. Lett.* **2012**, *45*, 51–62. [CrossRef]
97. Tahoun, I.F.; Gab-Allah, M.A.; Yamani, R.N.; Shehata, A.B. Development and validation of a reliable LC-MS/MS method for simultaneous determination of deoxynivalenol and T-2 toxin in maize and oats. *Microchem. J.* **2021**, *169*, 106599. [CrossRef]
98. Arroyo-Manzanares, N.; De Ruyck, K.; Uka, V.; Gámiz-Gracia, L.; García-Campaña, A.M.; De Saeger, S.; Diana Di Mavungu, J. In-house validation of a rapid and efficient procedure for simultaneous determination of ergot alkaloids and other mycotoxins in wheat and maize. *Anal. Bioanal. Chem.* **2018**, *410*, 5567–5581. [CrossRef]
99. Rahmani, A.; Jinap, S.; Khatib, A.; Tan, C.P. Simultaneous determination of aflatoxins, ochratoxin A, and zearalenone in cereals using a validated RP-HPLC method and PHRED derivatization system. *J. Liq. Chromatogr. Relat. Technol.* **2013**, *36*, 600–617. [CrossRef]
100. Pietsch, C.; Kersten, S.; Burkhardt-Holm, P.; Valenta, H.; Dänicke, S. Occurrence of deoxynivalenol and zearalenone in commercial fish feed: An initial study. *Toxins* **2013**, *5*, 184–192. [CrossRef]
101. Albero, B.; Fernández-Cruz, M.L.; Pérez, R.A. Simultaneous Determination of 15 Mycotoxins in Aquaculture Feed by Liquid Chromatography–Tandem Mass Spectrometry. *Toxins* **2022**, *14*, 316. [CrossRef]
102. Li, P.; Zhang, Q.; Zhang, W. Immunoassays for aflatoxins. *TrAC Trends Anal. Chem.* **2009**, *28*, 1115–1126. [CrossRef]
103. Meulenberg, E.P. Immunochemical methods for ochratoxin A detection: A review. *Toxins* **2012**, *4*, 244–266. [CrossRef]
104. Sanders, M.; Guo, Y.; Iyer, A.; Ruiz García, Y.; Galvita, A.; Heyerick, A.; Deforce, D.; Risseeuw, M.D.P.; Van Calenbergh, S.; Bracke, M. An immunogen synthesis strategy for the development of specific anti-deoxynivalenol monoclonal antibodies. *Food Addit. Contam. Part A* **2014**, *31*, 1751–1759. [CrossRef]
105. Li, S.; Wang, J.; Sheng, W.; Wen, W.; Gu, Y.; Wang, S. Fluorometric lateral flow immunochromatographic zearalenone assay by exploiting a quencher system composed of carbon dots and silver nanoparticles. *Microchim. Acta* **2018**, *185*, 388. [CrossRef]
106. Ezquerra, A.; Vidal, J.C.; Bonel, L.; Castillo, J.R. A validated multi-channel electrochemical immunoassay for rapid fumonisin B1 determination in cereal samples. *Anal. Methods* **2015**, *7*, 3742–3749. [CrossRef]
107. Oswald, S.; Karsunke, X.Y.Z.; Dietrich, R.; Märtlbauer, E.; Niessner, R.; Knopp, D. Automated regenerable microarray-based immunoassay for rapid parallel quantification of mycotoxins in cereals. *Anal. Bioanal. Chem.* **2013**, *405*, 6405–6415. [CrossRef]
108. McNamee, S.E.; Bravin, F.; Rosar, G.; Elliott, C.T.; Campbell, K. Development of a nanoarray capable of the rapid and simultaneous detection of zearalenone, T2-toxin and fumonisin. *Talanta* **2017**, *164*, 368–376. [CrossRef]
109. D'Agnello, P.; Vita, V.; Franchino, C.; Urbano, L.; Curiale, A.; Debegnach, F.; Iammarino, M.; Marchesani, G.; Chiaravalle, A.E.; De Pace, R. ELISA and UPLC/FLD as Screening and Confirmatory Techniques for T-2/HT-2 Mycotoxin Determination in Cereals. *Appl. Sci.* **2021**, *11*, 1688. [CrossRef]
110. Peltomaa, R.; Abbas, A.; Yli-Mattila, T.; Lamminmäki, U. Single-step noncompetitive immunocomplex immunoassay for rapid aflatoxin detection. *Food Chem.* **2022**, *392*, 133287. [CrossRef]
111. Wang, Y.-K.; Yan, Y.-X.; Li, S.-Q.; Wang, H.; Ji, W.-H.; Sun, J.-H. Simultaneous quantitative determination of multiple mycotoxins in cereal and feedstuff samples by a suspension array immunoassay. *J. Agric. Food Chem.* **2013**, *61*, 10948–10953. [CrossRef]
112. Vidal, J.C.; Bertolín, J.R.; Ezquerra, A.; Hernández, S.; Castillo, J.R. Rapid simultaneous extraction and magnetic particle-based enzyme immunoassay for the parallel determination of ochratoxin A, fumonisin B1 and deoxynivalenol mycotoxins in cereal samples. *Anal. Methods* **2017**, *9*, 3602–3611. [CrossRef]
113. Bian, Y.; Huang, X.; Ren, J. Sensitive and homogenous immunoassay of fumonisin in foods using single molecule fluorescence correlation spectroscopy. *Anal. Methods* **2016**, *8*, 1333–1338. [CrossRef]

114. Beloglazova, N.V.; Foubert, A.; Gordienko, A.; Tessier, M.D.; Aubert, T.; Drijvers, E.; Goryacheva, I.; Hens, Z.; De Saeger, S. Sensitive QD@ SiO2-based immunoassay for triplex determination of cereal-borne mycotoxins. *Talanta* **2016**, *160*, 66–71. [CrossRef] [PubMed]
115. Beloglazova, N.V.; Graniczkowska, K.; Njumbe Ediage, E.; Averkieva, O.; De Saeger, S. Sensitive flow-through immunoassay for rapid multiplex determination of cereal-borne mycotoxins in feed and feed ingredients. *J. Agric. Food Chem.* **2017**, *65*, 7131–7137. [CrossRef]
116. He, Q.-H.; Xu, Y.; Wang, D.; Kang, M.; Huang, Z.-B.; Li, Y.-P. Simultaneous multiresidue determination of mycotoxins in cereal samples by polyvinylidene fluoride membrane based dot immunoassay. *Food Chem.* **2012**, *134*, 507–512. [CrossRef]
117. Wang, X.; He, Q.; Xu, Y.; Liu, X.; Shu, M.; Tu, Z.; Li, Y.; Wang, W.; Cao, D. Anti-idiotypic VHH phage display-mediated immuno-PCR for ultrasensitive determination of mycotoxin zearalenone in cereals. *Talanta* **2016**, *147*, 410–415. [CrossRef]
118. Beloglazova, N.; Lenain, P.; Tessier, M.; Goryacheva, I.; Hens, Z.; De Saeger, S. Bioimprinting for multiplex luminescent detection of deoxynivalenol and zearalenone. *Talanta* **2019**, *192*, 169–174.
119. Ye, J.; Zheng, M.; Ma, H.; Xuan, Z.; Tian, W.; Liu, H.; Wang, S.; Zhang, Y. Development and Validation of an Automated Magneto-Controlled Pretreatment for Chromatography-Free Detection of Aflatoxin B1 in Cereals and Oils through Atomic Absorption Spectroscopy. *Toxins* **2022**, *14*, 454. [CrossRef]
120. Li, L.; Ren, S.; Shao, M.; De Saeger, S.; Song, S.; Yan, L. A competitive immunoassay for zearalenone with integrated poly (dimethysiloxane) based microarray assay. *Anal. Methods* **2018**, *10*, 4036–4043. [CrossRef]
121. Beloglazova, N.V.; Speranskaya, E.S.; Wu, A.; Wang, Z.; Sanders, M.; Goftman, V.V.; Zhang, D.; Goryacheva, I.Y.; De Saeger, S. Novel multiplex fluorescent immunoassays based on quantum dot nanolabels for mycotoxins determination. *Biosens. Bioelectron.* **2014**, *62*, 59–65. [CrossRef]
122. Xu, L.; Zhang, Z.; Zhang, Q.; Zhang, W.; Yu, L.; Wang, D.; Li, H.; Li, P. An on-site simultaneous semi-quantification of aflatoxin B1, zearalenone, and T-2 toxin in maize-and cereal-based feed via multicolor immunochromatographic assay. *Toxins* **2018**, *10*, 87. [CrossRef]
123. Li, S.; Sheng, W.; Wen, W.; Gu, Y.; Wang, J.; Wang, S. Three kinds of lateral flow immunochromatographic assays based on the use of nanoparticle labels for fluorometric determination of zearalenone. *Microchim. Acta* **2018**, *185*, 238. [CrossRef]
124. Majer-Baranyi, K.; Adányi, N.; Székács, A. Biosensors for Deoxynivalenol and Zearalenone Determination in Feed Quality Control. *Toxins* **2021**, *13*, 499. [CrossRef]
125. Li, R.; Wen, Y.; Wang, F.; He, P. Recent advances in immunoassays and biosensors for mycotoxins detection in feedstuffs and foods. *J. Anim. Sci. Biotechnol.* **2021**, *12*, 108. [CrossRef]
126. Maragos, C. Biosensors for mycotoxin analysis: Recent developments and future prospects. *World Mycotoxin J.* **2009**, *2*, 221–238. [CrossRef]
127. Tan, H.; Guo, T.; Zhou, H.; Dai, H.; Yu, Y.; Zhu, H.; Wang, H.; Fu, Y.; Zhang, Y.; Ma, L. A simple mesoporous silica nanoparticle-based fluorescence aptasensor for the detection of zearalenone in grain and cereal products. *Anal. Bioanal. Chem.* **2020**, *412*, 5627–5635. [CrossRef]
128. Llorent-Martínez, E.J.; Fernández-Poyatos, M.P.; Ruiz-Medina, A. Automated fluorimetric sensor for the determination of zearalenone mycotoxin in maize and cereals feedstuff. *Talanta* **2019**, *191*, 89–93. [CrossRef]
129. Anastasiadis, V.; Raptis, I.; Economou, A.; Kakabakos, S.E.; Petrou, P.S. Fast Deoxynivalenol Determination in Cereals Using a White Light Reflectance Spectroscopy Immunosensor. *Biosensors* **2020**, *10*, 154. [CrossRef]
130. Yin, L.; You, T.; El-Seedi, H.R.; El-Garawani, I.M.; Guo, Z.; Zou, X.; Cai, J. Rapid and sensitive detection of zearalenone in corn using SERS-based lateral flow immunosensor. *Food Chem.* **2022**, *396*, 133707. [CrossRef]
131. Qing, Y.; Li, X.; Chen, S.; Zhou, X.; Luo, M.; Xu, X.; Li, C.; Qiu, J. Differential pulse voltammetric ochratoxin A assay based on the use of an aptamer and hybridization chain reaction. *Microchim. Acta* **2017**, *184*, 863–870. [CrossRef]
132. Liu, Y.; Chen, Y.; Xu, W.; Song, D.; Han, X.; Long, F. Rapid, Sensitive On-Site Detection of Deoxynivalenol in Cereals Using Portable and Reusable Evanescent Wave Optofluidic Immunosensor. *Int. J. Environ. Res. Public Health* **2022**, *19*, 3759. [CrossRef] [PubMed]
133. McMullin, D.; Mizaikoff, B.; Krska, R. Advancements in IR spectroscopic approaches for the determination of fungal derived contaminations in food crops. *Anal. Bioanal. Chem.* **2015**, *407*, 653–660. [CrossRef] [PubMed]
134. Pojić, M.M.; Mastilović, J.S. Near infrared spectroscopy—Advanced analytical tool in wheat breeding, trade, and processing. *Food Bioprocess Technol.* **2013**, *6*, 330–352. [CrossRef]
135. Cen, H.; He, Y. Theory and application of near infrared reflectance spectroscopy in determination of food quality. *Trends Food Sci. Technol.* **2007**, *18*, 72–83. [CrossRef]
136. Deng, J.; Jiang, H.; Chen, Q. Characteristic wavelengths optimization improved the predictive performance of near-infrared spectroscopy models for determination of aflatoxin B1 in maize. *J. Cereal Sci.* **2022**, *105*, 103474. [CrossRef]
137. Ning, H.; Wang, J.; Jiang, H.; Chen, Q. Quantitative detection of zearalenone in wheat grains based on near-infrared spectroscopy. *Spectrochim. Acta Part A Mol. Biomol. Spectrosc.* **2022**, *280*, 121545. [CrossRef]

138. Tyska, D.; Mallmann, A.O.; Vidal, J.K.; de Almeida, C.A.A.; Gressler, L.T.; Mallmann, C.A. Multivariate method for prediction of fumonisins B1 and B2 and zearalenone in Brazilian maize using Near Infrared Spectroscopy (NIR). *PLoS ONE* **2021**, *16*, e0244957. [CrossRef]
139. Tyska, D.; Mallmann, A.; Gressler, L.T.; Mallmann, C.A. Near-infrared spectroscopy as a tool for rapid screening of deoxynivalenol in wheat flour and its applicability in the industry. *Food Addit. Contam. Part A* **2021**, *38*, 1958–1968. [CrossRef]

molecules

MDPI

Article

Comprehensive Characterization of Triterpene Saponins in Rhizoma Panacis Japonici by Offline Two-Dimensional Liquid Chromatography Coupled to Quadrupole Time-of-Flight Mass Spectrometry

Subinuer Yasen, Chengrui Li, Siyuan Wang, Yixin Dong, Hang Li, Jie Chen, Yifan Meng, Ping Yu and Haiyan Zou *

School of Traditional Chinese Medicine, Capital Medical University, Beijing 100069, China
* Correspondence: zouhy@ccmu.edu.cn

Abstract: Rhizoma Panacis Japonici (RPJ) is an ancient herbal medicine from China that has long been employed for its medicinal benefits in relieving arthritis physical debility and diverse afflictions. The primary bioactive constituents found in RPJ are triterpene saponins, which exhibit numerous pharmacological actions, including anti-inflammatory, antioxidant, and immunomodulating effects. The present study established a straightforward and effective approach for characterizing triterpene saponins in RPJ. An offline HILIC × RP LC/QTOF-MS method was developed, along with a self-constructed in-house database containing 612 saponins reported in the Panax genus and 228 predicted metabolites. The approach achieved good chromatographic performance in isolating triterpene saponins of RPJ, with the HILIC column as the first dimension (^{1}D) and the BEH C18 column as the second dimension (^{2}D). The developed two-dimensional liquid chromatography system exhibited an orthogonality of 0.61 and a peak capacity of 1249. Detection was performed using a QTOF mass spectrometer in a data-independent manner (MS^E) in a negative ion mode. Using the in-house database, the collected MS data were processed by an automatic workflow on UNIFI 1.8.2 software, which included data correction, matching of precursor and product ions, and peak annotation. In this study, 307 saponins were characterized from RPJ and 76 saponins were identified for the first time in *Panax japonicus*. This research not only enhances our understanding of the chemical characteristics of RPJ but also offers a simple and efficient method for analyzing the complex composition of herbal medicine.

Keywords: *Panax japonicus*; triterpene saponin; offline two-dimensional liquid chromatography; quadrupole time-of-flight mass spectrometry; in-house database; structural characterization

1. Introduction

Herbal medicines (HM) have been employed in traditional medical systems across the globe for several centuries and have garnered growing acknowledgment for their therapeutic properties. These medicines typically consist of complex chemical components [1]. Elucidating the compounds present in an herb is a fundamental question to ensure its efficacy and safety in traditional Chinese medicine research [2]. With advancements in analytical technology, particularly chromatography combined with high-resolution mass spectrometry (HRMS), significant progress has been made in characterizing the chemical composition of herbal medicine. However, it is imperative to recognize that the scientific community continues to face difficulties in effectively segregating and characterizing the intricate chemical components present in HM [3].

Liquid chromatography-HRMS (LC-HRMS) is the most widely used technique for characterizing metabolites in HM, especially small metabolites. Nevertheless, the significant variations in polarity, structures, and content of chemical constituents in HM present a growing challenge for one-dimensional (^{1}D) LC-MS. The peak capacity attainable with a single column or separation mechanism was insufficient for separating complex samples, making it hard to obtain MS^n information on minor or trace components. Moreover, the interference from coeluting components increases the difficulty in structural identification or even leads to reproducible results [4]. Therefore, enhancing separation capabilities and reducing the coelution fraction are essential for improving the sensitivity and reproducibility of LC-MS. Recently, two-dimensional (2D) LC-HRMS has emerged as a powerful tool for profiling complex chemical systems, such as natural products, biosamples, environmental contaminants and food additives [5–7]. 2D LC significantly increases peak capacities to several thousand or even over 10,000 by connecting two columns with different separation mechanisms in series, such as the combination of normal-phase (NP) and reversed-phase (RP), ion exchange (IE) and RP, as well as HILIC and RP [5]. Shi Qiu et al. [8] developed an offline 2D LC-MS system to analyze ginsenosides in *P. ginseng* leaves. They used HILIC-HPLC as the first dimension (^{1}D) to isolate the extract into multiple fractions. After concentrated, each fraction was analyzed with the second dimension (^{2}D) using RP-UHPLC/LTQ-Orbitrap MS instrument. The system showed a practical peak capacity of 11,000, leading to the characterization of 646 ginsenosides, a five-fold increase compared to those identified using only RP-UHPLC/LTQ-Orbitrap-MS. 2D LC can also be operated in online mode using special instruments that facilitate the continuous transfer of fractions from the ^{1}D column to the ^{2}D column. Online mode offers increased automation and throughput compared to offline mode. However, in addition to equipment requirements, the chromatographic conditions in the online mode are usually difficult to optimize due to the compatibility of mobile phases between the ^{1}D and ^{2}D isolation [9]. On the other hand, offline 2D LC provides flexibility, ease of operation, and the potential to perform multidimensional LC separations [10].

HRMS provides extensive coverage of chemicals, a broad dynamic range, precise mass measurements, and distinguishable isotope distributions [11]. As a result, it has been effectively utilized for untargeted analysis in various research fields, including natural product [12], proteomics [13], foodomics [6], lipidomics [5], etc. Data-dependent (DDA) and data-independent (DIA) acquisition modes are the primary data acquisition modes for untargeted metabolite analysis. The DDA mode is valuable for linking MS^2 spectra to the precursor ions and creating MS^1-MS^2 datasets, but compounds with low MS abundance cannot undergo fragmentation. On the other hand, DIA modes, such as SWATH and MS^E, can continuously and impartially capture MS^2 information for all compounds, resulting in notably greater spectral coverage compared to DDA modes [11,14]. So, some studies have combined the two modes for improving accuracy and coverage of the structural identification in profiling chemical constituents of complex sample [15,16].

Efficient management of massive MS/MS data of 2D LC-HRMS is essential for in-depth chemical characterization of complex sample. Computer-aided database searches are commonly used for untargeted analysis. Various software and algorithms are employed for automatic processing of MS data, such as UNIFI 1.8.2 software (Waters) [17], ACD/MS Structure ID Suite (ACD/Labs) [18], as well as MS-DIAL [19], XCMS [20], and Open MS 2.0 [21]. A comprehensive database specific to the samples being analyzed is important for the identification of metabolites. While online databases like MassBank and Metline are available, additional prediction strategies have been developed to enhance database coverage. In the case of saponins, structure predictions primarily focus on substitution patterns, such as acetylation, formylation, malonylation, and types of sugar substituents. By utilizing this approach, 945 saponins, including 662 potentially novel ginsenosides, were identified from the leaves of *P. notoginseng* using UNIFI software combined with an in-house database [22].

Rhizoma Panacis Japonici (RPJ) derives from *Panax japonicus* C. A. Mey., a species of the *Araliaceae* family [23]. It has been used as a folk medicine in China for over 200 years and has been recorded in the Chinese Pharmacopoeia since 1977. RPJ and its preparations are primarily utilized in the clinical treatment of rheumatoid arthritis [24]. RPJ contains triterpenoid saponins, polysaccharides, minerals, and amino acids. Of these components, triterpene saponins are considered the primary active ingredients in RPJ, display a diverse array of pharmacological properties, including anti-inflammatory, antioxidant, and anti-myocardial ischemia effects [25,26]. We were the first to report the therapeutic effects of the total saponin from RPJ on experimental autoimmune encephalomyelitis, a classical animal model of multiple sclerosis [27]. A Chinese national invention patent (No. zl201410041725.5) has been granted for this discovery [28]. Therefore, elucidating the saponin constituents in RPJ is essential for further research.

To date, about 113 triterpene saponins have been isolated from RPJ, which can be categorized into the protopanaxdiol (PPD), protopanaxtriol (PPT), octillol (OT), oleanolic acid (OA), and ursonic acid (UA) types [29,30]. Unlike *P. ginseng* and *P. notoginseng*, which are well-known medicinal plants in the *P.* genus, RPJ is unique for its high content of OA-type saponins along with a small amount of dammarane-type saponins. In previous studies, RPJ has been analyzed using different methods, such as UHPLC-Q-Exactive Orbitrap HRMS and UFLC-Triple TOF-MS/MS, resulting in the characterization of 53 and 82 saponins, respectively [31,32]. Additionally, Chunxia Zhang et al. [15] expanded on this research by identifying 178 components in RPJ using RP LC/IM-QTOF-MS combined with data-dependent and data-independent acquisition strategies. However, there is still a gap in characterizing saponins of RPJ. To address this, we established an offline HILIC × RP LC/QTOF-MS system. Additionally, we developed a comprehensive in-house database that documents 612 saponins found in the *P.* genus and 228 predicted metabolites for characterizing saponins in RPJ. The workflow is illustrated in Figure 1. The RPJ was extracted and separated into multiple fractions using a HILIC column (^{1}D HPLC). Each fraction was then further separated by a BEH-C18 column (^{2}D UPLC) and detected using QTOF-MS/MS in negative mode. The mass data were efficiently managed, and the saponins present in RPJ were automatically identified with UNIFI software. Interpretations were made by studying the fragmentation behaviors of 23 reference saponins (Figure 2).

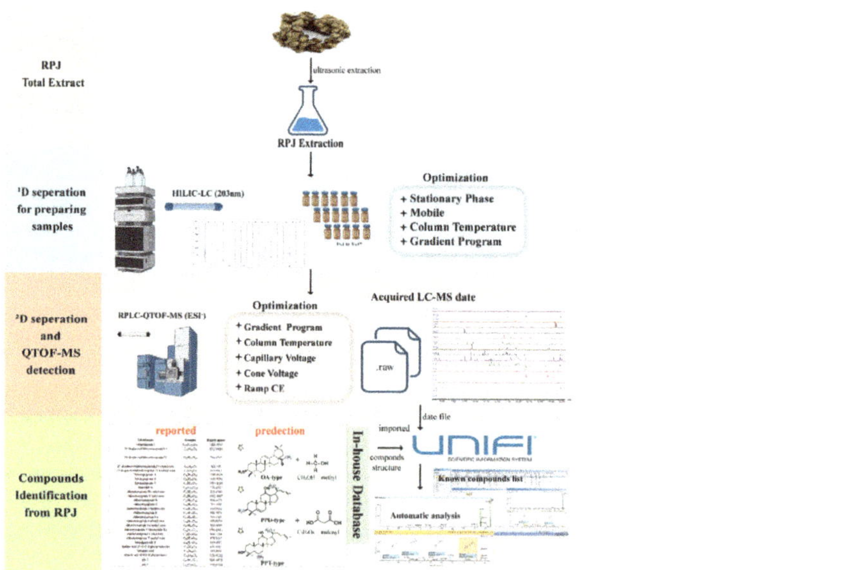

Figure 1. A comprehensive workflow for profiling triterpene saponins in Rhizoma Panacis Japonici.

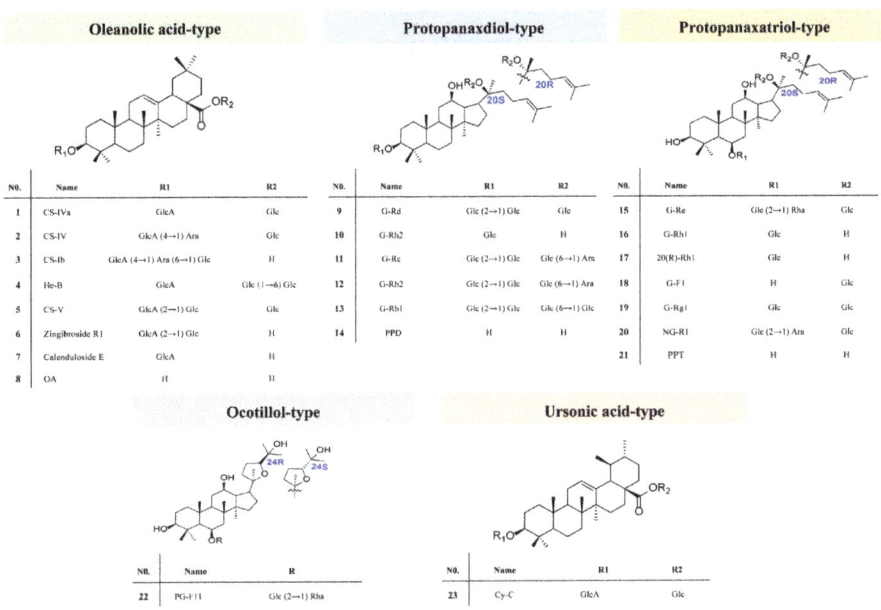

N0.	Name	R1	R2
1	CS-IVa	GlcA	Glc
2	CS-IV	GlcA (4→1) Ara	Glc
3	CS-Ib	GlcA (4→1) Ara (6→1) Glc	H
4	He-B	GlcA	Glc (1→6) Glc
5	CS-V	GlcA (2→1) Glc	Glc
6	Zingibroside R1	GlcA (2→1) Glc	H
7	Calenduloside E	GlcA	H
8	OA	H	H

N0.	Name	R1	R2
9	G-Rd	Glc (2→1) Glc	Glc
10	G-Rh2	Glc	H
11	G-Rc	Glc (2→1) Glc	Glc (6→1) Ara
12	G-Rb2	Glc (2→1) Glc	Glc (6→1) Ara
13	G-Rb1	Glc (2→1) Glc	Glc (6→1) Glc
14	PPD	H	H

N0.	Name	R1	R2
15	G-Re	Glc (2→1) Rha	Glc
16	G-Rh1	Glc	H
17	20(R)-Rh1	Glc	H
18	G-F1	H	Glc
19	G-Rg1	Glc	Glc
20	NG-R1	Glc (2→1) Ara	Glc
21	PPT	H	H

N0.	Name	R
22	PG-F11	Glc (2→1) Rha

N0.	Name	R1	R2
23	Cy-C	GlcA	Glc

Figure 2. The structures of 23 reference saponins used in the study: G, Ginsenoside; NG, Notoginsenoside; CS, chikusetsusaponin; Cy-C, Cynarasaponin C; He-B, hemsgiganoside B; PG-F11, Pseudoginsenoside F11; Glc, glucose; Rha, rhamnose; Ara, arabinose; Xyl, xylose; GlcA, glucuronic acid.

2. Results and Discussion

2.1. Optimization of the ^{1}D and ^{2}D-LC Conditions

To isolate saponins from RPJ, we selected combinations of HILIC × RP to isolate triterpene saponins of RPJ according to some reports [8,15]. The ^{1}D (HILIC) and ^{2}D (RP) LC conditions were optimized systematically. For ^{1}D HPLC separations, we compared the separation performance of two stationary phases, including silica (Atlantis HILIC column) and amide (XBridge Amide column and BEH Amide column). As shown in Figure S1, columns packed with amide material performed better in retaining saponins of RPJ than the Atlantis HILIC column (silica). Compared with the BEH Amide column, more peaks with symmetrical peak shapes were observed when using the XBridge Amide column. Therefore, a XBridge Amide column was used for the ^{1}D-LC isolation of RPJ extract. Given that RPJ primarily contains acidic saponins (OA type), we evaluated the impact of water and additive of 0.1% formic acid (FA), 0.1 M ammonium formate (AF), and 0.1% trifluoroacetic acid (TFA) on the separation of saponins in RPJ using acetonitrile (CH_3CN) as the organic phase (Figure S2). Based on the number of peaks and peak symmetry, it was found that the addition of 0.1% formic acid yielded better results compared to 0.1 M AF and 0.1% TFA in separating the RPJ extract. Compared to using water alone, a greater number of peaks were observed within 19–23 min. Thus, 0.1% FA-CH_3CN was used as the mobile phase. Furthermore, we compared the impact of varying column temperatures (25–40 °C) on the separation of RPJ (Figure S3). The results showed that increasing the column temperature had a minimal impact on the resolution, so it was set at 30 °C.

For the ^{2}D-LC isolation, the performance of five different RP columns was evaluated in isolating RPJ extract and 23 reference saponins using an UPLC-QTOF-MS instrument. Based on previous research on ginsenoside [22,32–35], Scepter C18-120, CORTECS C18, BEH Shield RP18, HSS T3, and BEH C18 columns were selected. As depicted in Figure S4, the RPJ extract exhibited the highest number of chromatography peaks with the Scepter C18-120 column (298 peaks), followed by the BEH C18 column (287 peaks) and CORTECS- C18 column (257 peaks). Compared with others, the BEH C18 column was able to completely separate the 23 reference saponins and exhibited the best resolution of neighboring peaks. Due to the specificity of saponins, the BEH C18 column was selected for the ^{2}D UPLC separation. We then regulated the column temperatures and observed that enhancing the resolution of some minor peaks at 40 °C compared to 30 °C or 35 °C (Figure S5). Thus, the column temperature of the ^{2}D UPLC isolation was set at 40 °C.

2.2. Optimization of QTOF-MS Parameters

The key parameters of the Synapt^{TM} mass spectrometer (Waters, Milford, MA, USA) were fine-tuned to obtain maximum sensitivity and product ion information of saponins in RPJ. Firstly, both the positive and negative ion mode were utilized to analyze the reference saponins. Abundant fragments could be generated from the parent ion in negative ion mode as a result of consecutive neutral loss of the external sugar substituent. In contrast, adduct ions ($[M + Na]^+$ and $[M + NH_4]^+$) were produced in positive ion mode, and fewer fragments were apparent in MS^2 spectra. The results were consistent with the previously described [36]. Thus, we chose negative mode in this study. Next, the capillary voltage and cone voltage were tested by evaluating the intensity of four types of saponins, including chikusetsusaponin V (OA type), ginsenoside Rb1 (PPD type), ginsenoside Rh1 (PPT type) and Pseudoginsenoside F11 (OT type). For the capillary voltage (Figure 3a), the adduct ion intensity of target compounds was differently changed at 1.0–3.0 kV, but the ionization was relatively high at 1.5 kV. Thus, the capillary voltage was set at 1.5 kV. The cone voltage could induce in-source dissociation of saponins, and an optimal value would enable a higher response and detection sensitivity. All index saponins produced relatively high ionic intensity of $[M - H]^-$ at 40 V, indicating that 40 V was the optimal cone voltage (Figure 3b).

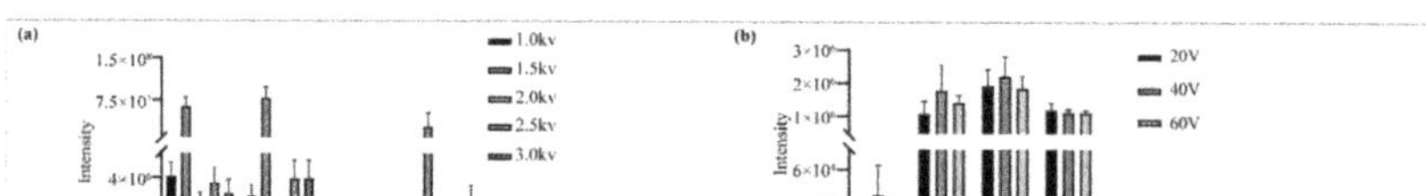

Figure 3. Influence of (**a**) capillary voltage, (**b**) cone voltage, and (**c**) collision energy on MS^2 behaviors of the four representative saponins.

Collision energy (CE) is vital to induce dissociation of protonated or deprotonated parent ions [37]. In MS, deglycosylated fragments and sapogenin ions are the main diagnostic ions of saponins. Specifically, we selected G-Rh1, PG-F11, CS-V, and G-Rb1 as representative saponins with one to four sugar units, respectively. Optimization of CE was carried out within specific energy ranges: 20–40 eV, 30–50 eV, 40–60 eV, 50–70 eV, and 60–80 eV. As shown in Figure 3c. The energy required to generate sapogenin ions increased with the number of glycosidic bonds to some extent. G-Rh1, a monoglycoside, produced relatively high levels of *m*/*z* 475.3764 [sapogenin − H]$^-$ and its fragments *m*/*z* 391.2820 at lower energy (CE 30–50 eV). The optimal CE for PG-F11 (diglycoside) and CS-V (triglycoside) was 40–60 eV and 50–70 eV, respectively. In contrast, G-Rb1 (tetraglycoside) generated *m*/*z* 475.3764 [sapogenin − H]$^-$ at higher collision energies of 50–70 eV and 60–80 eV. Notably, at CE 50–70 eV, the mass spectra of the four saponins showed a variety of fragments and relatively high levels of aglycone ions. Therefore, the collision energy was set at 50–70 eV in the MS measurement of the sample.

2.3. Evaluation and Method Validation

The assessment of the offline 2D LC system's separation performance involved orthogonality and peak capacity. By applying asterisk Equations [38], the spreading of 23 reference saponins was calculated, resulting in an orthogonality value of 0.61 (Figure 4). The parameters of the four crossing lines were calculated at 0.92 (Z_-), 0.56 (Z_+), 0.98 (Z_1) and 0.72 (Z_2), respectively. The ^{1}D and ^{2}D-LC exhibited a peak capacity of 97 (mean peak width 0.33 min) and 135 (mean peak width 0.27 min), respectively. Consequently, the 2D LC system exhibited a theoretical peak capacity of 13,175, with an effective peak capacity of 1249. These results indicated that the developed offline HILIC × RP LC system significantly enhanced the resolution of saponins in RPJ. For example (Figure 5), the chromatographic peak at t_R 11.23 min of RPJ extract detected by RP-LC/QTOF-MS potentially contained four coeluted compounds (*m*/*z* 1169.6055, *m*/*z* 1005.5336, *m*/*z* 941.4769, *m*/*z* 887.4979). Characterizing their structures based on MS^2 data was challenging due to the mixed fragment ions of all the coeluted compounds. However, the developed 2D LC-MS system successfully distributed these saponins in Fr.14, 13, 10 and 7, and identified as 6-O-[β-D-glucopyranosyl-(1,2)-β-D-glucopyranosyl]-20-O-[β-D-glucopyranosyl-(1,4)-β-D-glucopyranosyl]-20(S)-protopanaxatriol or isomer, notoginsenoside G or isomer, (OA+O)-GlcA-Xyl-Glc and (PPT+O)-Glc-Rha-malonyl. This result highlights the significant advantage of the offline 2D LC-MS system in resolving coeluted and trace components in samples.

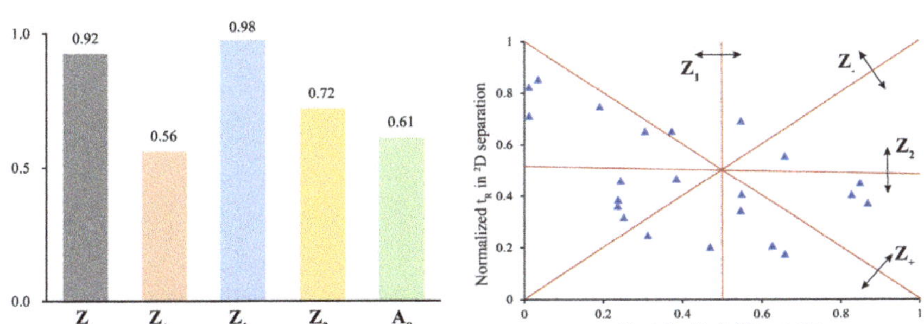

Figure 4. Orthogonality of the offline HILIC × RP LC system with asterisk equations using 23 reference saponins.

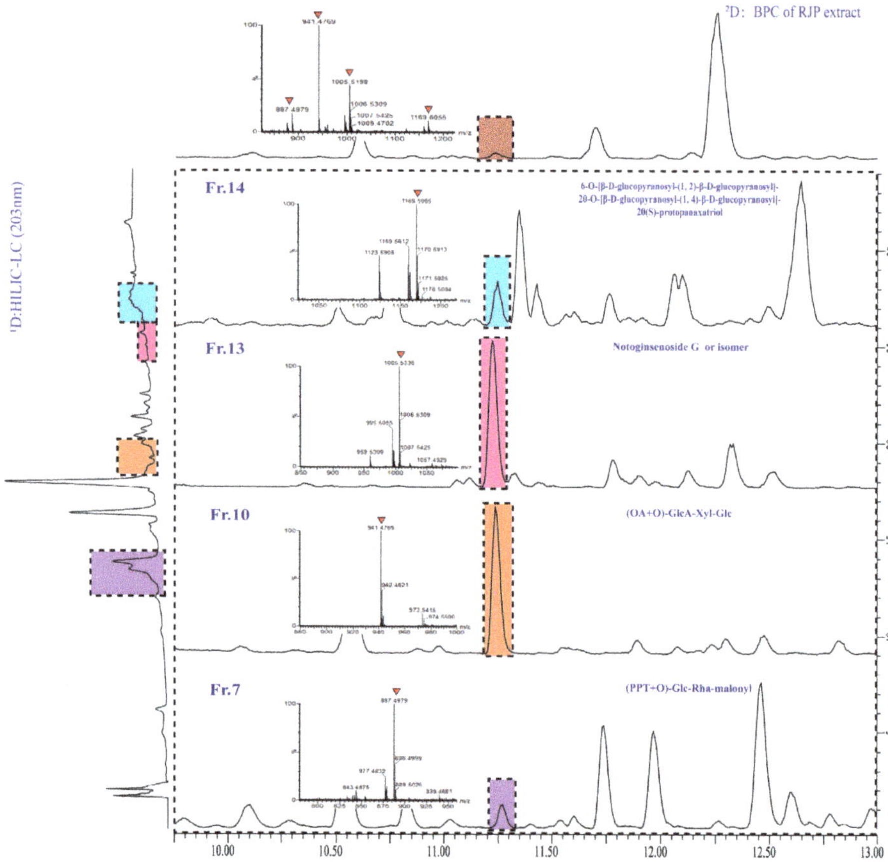

Figure 5. Enlarged ^{1}D HPLC-UV (203 nm)-^{2}D base peak ion (BPI) chromatograms of RPJ extract and selected fractions, showing the four coeluted saponins in RPJ extract were separated and detected in fractions 14, 13, 10, and 7.

Additionally, simplified method validation was conducted for both ^{1}D and ^{2}D separations as the reports [33,39], including repeatability, inter-/intra-day precision and limit of detection (LOD). Five index saponins (including G-Rh1, G-Re, CS-V, CS-IVa and CS-IV) were used to evaluate precision and repeatability. The relative standard deviation (RSD, %) for inter-/intra-day precision of ^{1}D and ^{2}D separation ranged from 0.52% to 4.03% and from 0.70% to 6.43% (Tables S1–S4), respectively. For repeatability of the offline 2D LC-MS method, the RSD of the five saponins ranged from 1.11% to 3.87% (Table S5). The LOD of G-Rh1, G-Re, CS-V, CS-IVa and CS-IV were 1.19 ng, 1.50 ng, 1.23 ng, 2.40 ng, and 1.20 ng, respectively. The findings suggested that the devised method of HILIC × RPLC/QTOF-MS is stable, sensitive and repeatable.

2.4. Systematic Characterization of the Triterpene Saponins in RPJ

The triterpenoid saponins of RPJ mainly consist of OA and dammarane types, which can be categorized into PPD, PPT, OT type, and varied C17 side chains. Regarding the sugar constituents, GlcA ($C_6H_{10}O_7$), Glc ($C_6H_{12}O_6$), Rha ($C_6H_{12}O_5$), Ara ($C_5H_{10}O_5$) and Xyl ($C_5H_{10}O_5$) have been reported in *P.* species [8,36], showing the neutral loss of 176.0319 Da, 162.0550 Da, 146.0542 Da, 132.0365 Da and 132.0365 Da, respectively. In the present study,

Xyl was used to address pentose residue for the neutral loss of 132.0365 Da. Furthermore, esterified or acylated saponins were also characterized in RPJ (Table S8).

2.4.1. OA-Type Saponins

150 OA-type saponins in RPJ were identified in this study. The major diagnostic fragments for these saponins were the dehydrogenated aglycone ion at *m/z* 455.3501 and the neutral loss of 43.9990 Da (CO_2). The sugar chains of OA-type saponins are typically attached at positions 28-COOH and/or 3-OH. It is observed that the glycoside linkage at C-28 is more susceptible to breakage compared to the one at the C-3 position in the negative ion mode. Thus, the substitution positions or isomers of various sugar chains in OA-type saponins can be determined by analyzing the relative abundance of deglycosylated fragments. For example, CS-V and He-B are isomers with two sugar chains. They both produced $[M - H]^-$ ion at *m/z* 955.4883 ($C_{48}H_{76}O_{19}$) and deprotonated sapogenin ion at *m/z* 455.3519 (Figure 6). CS-V (3-GlcA-Glc, 28-Glc) generated deglucose chain fragments at *m/z* 793.4343 ($[M - H - Glc]^-$), along with ions at *m/z* 731.4336 ($[M - H - Glc - H_2O - CO_2]^-$), 613.3718 ($[M - H - H_2O - 2Glc]^-$), and 569.3819 ($[M - H - 2Glc - CO_2 - H_2O]^-$). On the other hand, He-B produced a high-intensity $[M - H - 2Glc]^-$ ion at *m/z* 631.3838, indicating that two glucose molecules are linked to the C-28 position instead of C-3. Similar cleavage behaviors were also observed in compounds CS-IV and CS-Ib, which is consistent with a previous report [40].

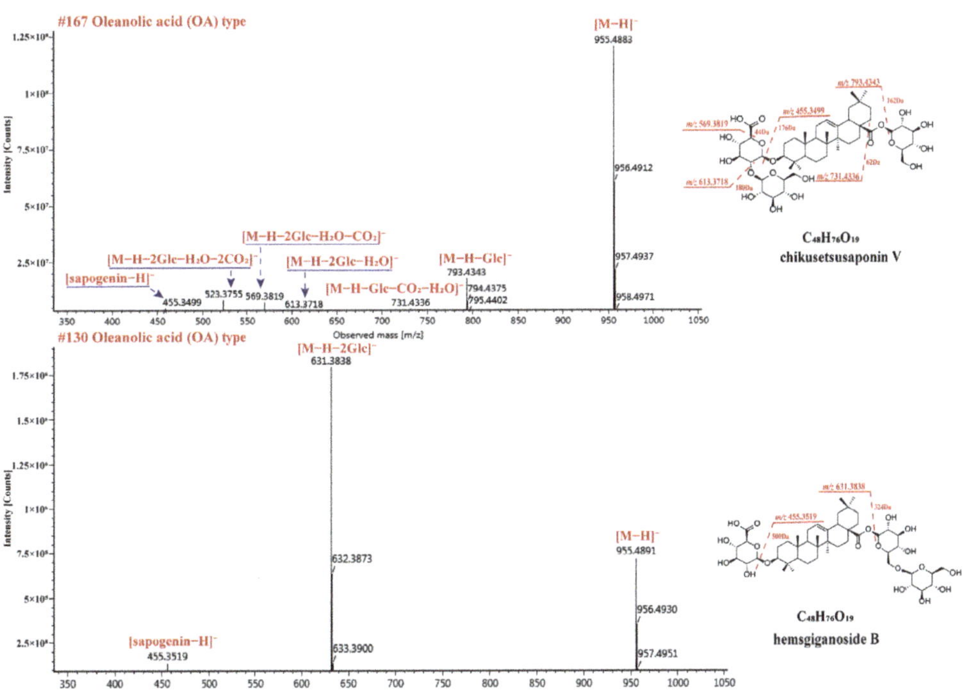

Figure 6. The MS^2 spectra for chikusetsusaponin V and hemsgiganoside B in negative ion mode.

It is worth noting that *P. japonicus* also contains UA-type saponins, which are aglycone isomers of the OA type, such as Cynarasaponin (Cy-C, UA-28-Glc-3-GlcA) and CS-IVa (OA-28-Glc-3-GlcA) [41]. They were distinguished by the retention time (CS-IVa, t_R, 21.72 min; Cy-C, t_R, 22.15 min) provided by the reference substances because of their highly similar MS^2 spectra. So, the sapogenin ion at *m/z* 455.3534 observed in MS^2 spectra was identified as the OA type. In MS/MS qualitative characterization, identifying high-level isomers has always been a challenging task especially in the absence of reference compounds. In recent years, techniques like energy-resolved (ER) MS and Ion Mobility (IM) MS have been utilized to distinguish isomers by detecting additional structural information, like optimal collision energy, half response collision energy, ion migration time, collision cross-section, and others. These techniques have been utilized for identifying isomers in HM without the need for reference compounds, such as lignan glycosides [39] and coumarins [42], among others. To our knowledge, the application of these techniques to distinguish UA and OA isomers has not yet been reported, which needs for further investigation.

2.4.2. Dammarane Type Saponins

PPT and PPD types are the main dammarane-type saponins in *P.* species. 31 PPD-type and 48 PPT-type saponins have been identified from RPJ in this study and they commonly produced $[M-H]^-$ and/or $[M+HCOO]^-$ in negative ion mode. The major diagnostic fragments of PPT and PPD-type were observed at m/z 475.3764 ($C_{30}H_{51}O_4$) → 391.2820 and 459.3860 ($C_{30}H_{51}O_3$) → 375.2902, respectively. Compound **67** (t_R, 11.26 min) gave $[M+HCOO]^-$ and $[M-H]^-$ ions at m/z 1169.5959 and 1123.5907, respectively, corresponding to the formula $C_{54}H_{92}O_{24}$ (Figure 7). The $[M-H]^-$ ion generated diagnostic fragments at m/z 475.3771 ($[M-H-4Glc]^-$) and 391.2861 ($[M-H-4Glc-C_6H_{12}]^-$), along with m/z 961.5367 ($[M-H-Glc]^-$), 799.4810 $[M-H-2Glc]^-$, 781.4718 ($[M-H-2Glc-H_2O]^-$) and 637.4304 ($[M-H-3Glc]^-$) ions. So, Compound **67** was characterized as 6-O-[β-D-glucopyranosyl-(1,2)-β-D-glucopyranosyl]-20-O-[β-D-glucopyranosyl-(1,4)-β-D-glucopyranosyl]-20(S)-protopanaxatriol. Compound **293** (t_R, 28.44 min), the molecular formula $C_{42}H_{72}O_{13}$ was confirmed by m/z 829.4951 ($[M+HCOO]^-$) and 783.4878 ($[M-H]^-$). The parent ion (m/z 783.4878) lost two Glc residues and gave m/z 459.3828 ($[\text{sapogenin}-H]^-$) and 375.2902 ($[\text{sapogenin}-H-C_6H_{12}]^-$) ions (Figure 7). Its molecular formula and cleavage behavior were consistent with ginsenoside Rg3 [43]. In addition, 16 OT-type saponins have been identified from RPJ and their characteristic ions were at m/z 491.3708 and 415.3257 in negative ion mode. Compound **15** (t_R, 5.47 min) showed abundant $[M+HCOO]^-$ and $[M-H]^-$ ions at m/z 1007.5425and 961.5380, corresponding to the formula $C_{48}H_{82}O_{19}$ (Figure 7). Fragment ions at m/z 815.4795 ($[M-H-Rha]^-$), 797.4684 ($[M-H-Rha-H_2O]^-$), 653.4269 ($[M-H-Rha-Glc]^-$), 635.4156 ($[M-H-Rha-Glc-H_2O]^-$), 491.3732 ($[M-H-Rha-2Glc]^-$), and 415.3229 ($[M-H-Rha-2Glc-C_3H_6O]^-$) were observed in MS^2 spectra. Thus, compound **15** was identified as octillol-Glc-Glc-Rha.

PPD or PPT-type saponins with dehydrated on sapogenin and varied C17 side chains have been reported from *P.* genus [29,44]. In the present study, 19 and 38 compounds of these types were characterized, which usually yield $[\text{sapogenin}-H]^-$ ion and specific fragments of the C17 side chain. For example, Compound **284** (t_R, 28.04 min) gave abundant ions at m/z 827.4825 ($[M+HCOO]^-$) and 781.4734 ($[M-H]^-$), which are consistent with the formula $C_{42}H_{70}O_{13}$. Its $[M-H]^-$ ion yielded fragments of m/z 619.4333 ($[M-H-Glc]^-$), 457.3685 ($[M-H-2Glc]^-$) and 373.2753 ($[M-H-2Glc-C_6H_{12}]^-$). The neutral losses of the sugar chain and C_6H_{12} indicated that compound **284** was ginsenoside 5-ene-PPD-Glc-Glc. Compound **31** (t_R, 7.78 min) was firstly identified in RPJ. It yielded m/z 861.4861 ($[M+HCOO]^-$) and 815.4783 ($[M-H]^-$) ions, indicating the formula $C_{42}H_{72}O_{15}$. The fragments m/z 669.4212, 507.3681, and 491.3747 were assigned to $[M-H-Rha]^-$, $[M-H-Rha-Glc]^-$, and $[M-H-Rha-Glc-H_2O]^-$ ions, respectively. The ion at m/z 507.3681 suggested two methyl substitutions in the side-chain of the aglycone (PPT). As reported in the literature [45], compound **31** was identified as floralquinquenoside B. Similar fragmentation behavior was also observed in compound **10** ($C_{41}H_{72}O_{15}$), which generated ions at m/z 803.4777 ($[M-H]^-$), 671.4326 ($[M-H-Xyl]^-$), 509.3869 ($[M-H-Xyl-Glc]^-$), and 391.2929 ($[M-H-Xyl-Glc-C_6H_{14}O_2]^-$). The fragments at m/z 509.3869 and m/z 391.2929 indicated that its aglycone was 23,24-OH-PPT. Therefore, compound **10** was characterized as 23,24-OH-PPT-Glc-Xyl.

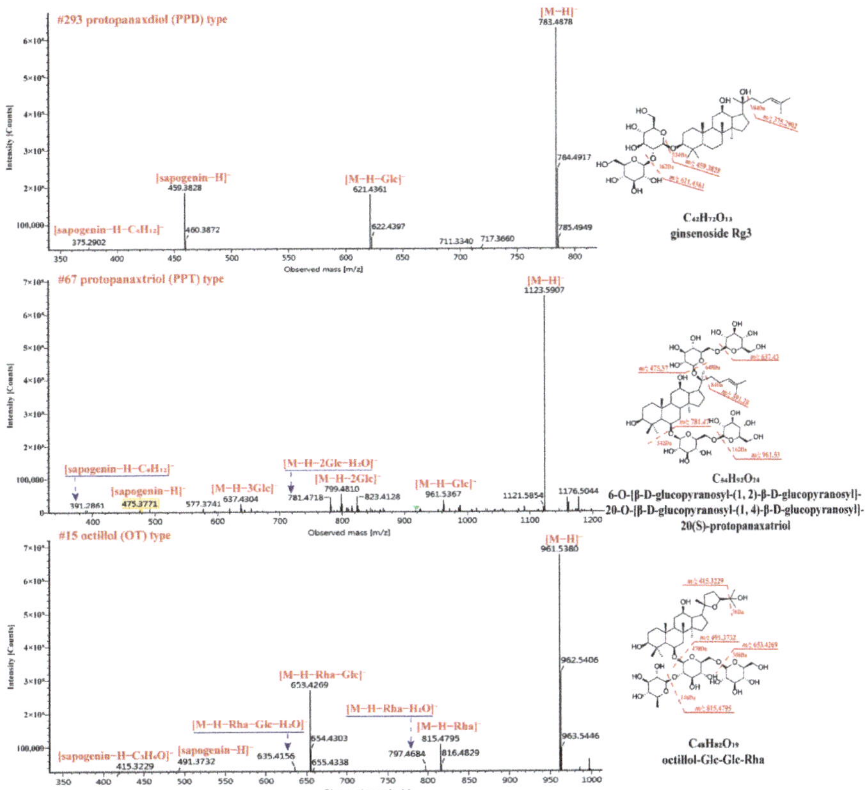

Figure 7. The MS^2 spectra for three dammarane-type saponins in negative ion mode.

2.4.3. Esterified and Acylated Type Saponins

Esterified and acylated saponins were also observed in RPJ. In the case of OA-type saponins, esterification could occur due to the presence of free carboxyl in sapogenin or glucuronic acid [30]. Among the OA-type saponins, 21 formylated, 7 ethylated, 1 polyacetylene (compound **269**), and 1 acylated (compound **255**) were found. The neutral losses of 13.9753 Da (methyl), 28.0311 Da (ethyl), 42.0113 Da (acetyl), 44.0104 Da (malonyl) could be observed in negative ion mode, respectively. Compound **270** (t_R, 27.36 min) gave [M + HCOO]$^-$ and [M − H]$^-$ ions at m/z 837.5613 and 791.4565, respectively, corresponding to the formula $C_{43}H_{68}O_{13}$ (Figure 8). The ion at m/z 763.4254 ([M − H − C_2H_4]$^-$), indicating the carboxyl was esterified. The precursor ion further generated m/z 631.3843 ([M − H − C_2H_4 − Ara]$^-$), 613.3740 ([M − H − C_2H_4 − Ara − H_2O]$^-$), 537.3583 ([M − H − C_2H_4 − Ara − $C_2H_4O_3$]$^-$), and 455.3347 ([M − H − C_2H_4 − Ara − GlcA]$^-$). Thus, compound **270** was identified as 28-desglucosylchikusetsusaponin IV ethyl ester.

On the other hand, dammarane-type saponins containing acetyl/malonyl substituents are also reported from *P.* species [8,36], and these substituents are usually at C20-sugar chain. We also found 2 malonylated and 16 acetylated saponins from RPJ in this study. In negative ion mode, these saponins exhibited [M − H − acetyl]$^-$ and [M − H − acetyl − H_2O]$^-$ diagnostic fragments, consistent with previous reports [36]. Compound 232 (t_R, 25.75 min, $C_{56}H_{94}O_{24}$) showed [M + HCOO]$^-$ and [M − H]$^-$ ions at m/z 1221.6252 and 1175.6190, respectively. In MS^2, the high intensity of ions at m/z 1107.5941 [M − H − acetyl]$^-$ and 1089.5822 [M − H − acetyl − H_2O]$^-$ were observed, accompany with fragments at m/z 945.5391 [M − H − acetyl − Glc]$^-$, 783.4871 [M − H − acetyl − 2Glc]$^-$,

621.4385 [M − H − acetyl − 3Glc]⁻, 603.4211 [M − H − acetyl − 3Glc − H_2O]⁻, and 459.3841 [M − H − acetyl − 4Glc]⁻. Compound 232 was identified as ginsenoside Ra6.

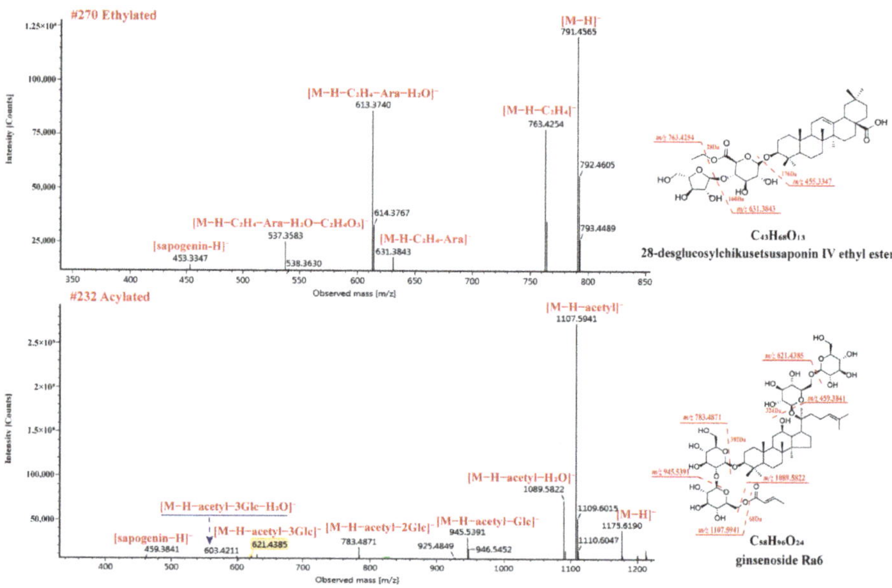

Figure 8. The MS^2 spectra for ethylated and acetylated saponins in negative ion mode.

In this study, 307 saponins were identified from RPJ using the developed 2D LC-QTOF-MS method based on an in-house database of *P.* genus. Among these saponins, 150 were categorized as OA-type saponins and their derivatives, while one was classified as a UA-type saponin. Furthermore, 156 dammarane-type saponins were identified, including C17 side-chain varied compounds along with esterified and acylated derivatives. In the chemical characterization of HM using 1D or 2D LC-HRMS, a comprehensive and specific database is necessary to enhance the efficiency of compound identification and ensure consistent results. Nonetheless, relying solely on computer-aided database searches may limit the exploration of new compounds to some extent.

3. Materials and Methods

3.1. Chemicals and Reagents

A total of 23 reference saponins were used in this study (Figure 2). Reference chikusetsusaponin-IVa, -IV, -Ib, -V, ginsenoside -Rd, -Rh2, -Re, -Rc, -Rb1, -Rb2, -F1, -Rg1, 20 (S) ginsenoside-Rh1, 20 (R) ginsenoside-Rh1, notoginsenoside-R1, zingibroside R1, Calenduloside E, oleanolic acid, protopanaxadiol, and protopanaxatriol were, respectively, purchased from Herbest Bio-Tech Co., Ltd. (Baoji, China), DeSiTe Biological Technology Co., Ltd. (Chengdu, China), Biopurify Phytochemicals Ltd. (Chengdu, China) and Pufei De Biotech Co., Ltd. (Chengdu, China) Hemsgiganoside B and cynarasaponin C were isolated from RPJ in our lab [41]. LC-grade acetonitrile and methanol were provided by Fisher Co. Ltd. (Emerson, IA, USA). The analytical grade trifluoroacetic acid, formic acid, and ammonium formate were purchased from Macklin Biotech Co., Ltd. (Shanghai, China). For the preparation of ultrapure water, a Milli-Q Reagent Water System (Millipore, Bedford, MA, USA) was utilized. The Rhizoma of *Panacis Japonici* was obtained from a pharmacy in Chengdu. It was identified as the rhizome of Panax japonicus C. A. Mey. by Associate Professor Li Jia of the College of Traditional Chinese Medicine at Capital Medical University. Voucher specimens with Batch No. PJ201901 have been deposited at the authors' lab in Capital Medical University (Beijing, China).

The columns used were as follows: XBridge Amide column (4.6 × 150 mm, 3.5 μm, Waters, USA) and (2.1 × 150 mm, 3.5 μm, Waters, USA), Atlantis HILIC silica column (2.1 × 150 mm, 3 μm, Waters, USA), BEH Amide column (2.1 × 100 mm, 1.7 μm, Waters), BEH C18 column (2.1 × 100 mm, 1.7 μm, Waters, USA), BEH Shield RP18 column (2.1 × 100 mm, 1.7 μm, Waters, USA), ACQUITY UPLC HSS T3 column (2.1 × 100 mm, 1.8 μm, Waters, USA), CORTECS UPLC C18 column (2.1 × 100 mm, 1.6 μm, Waters, USA) and Shim-pack Scepter C18-120 (2.1 × 150 mm, 1.9 μm, Shimadzu, Kyoto, Japan).

3.2. *Sample Preparation*

Ten g of fine powder of RPJ was extracted with the assistance of ultrasound (100 W, 50 Hz) for one hour at 30 °C using 200 mL of 70% methanol as solvent. The mixture was then centrifuged at 3000 rpm for 15 min. The supernatant liquid was filtered through a 0.22 μm PTFE filter membrane and stored at 4 °C for analysis.

3.3. *Offline HILIC×RP LC/QTOF-MS Conditions*

The first-dimensional (^{1}D) separation was performed on an Agilent 1200 HPLC system (Agilent Technologies, Santa Clara, CA, USA) using a Waters XBridge Amide column (4.6 × 150 mm, 3.5 μm). The mobile phase consisted of 0.1% FA (*v*/*v*) in water (A) and acetonitrile (CH_3CN, B), with the liner elution gradient: 0–3 min, 95% B; 3–5 min, 95–90% B; 5–14 min, 90–85% B; 14–17 min, 85% B; 17–22 min, 85–83% B; 22–27 min, 83–60% B; 27–32 min, 60% B; 32–35 min, 60–95% B. The column temperature was set at 30 °C and the wavelength was at 203 nm. The flow rate was 1.0 mL/min and the injection volume were 20 μL. Seventeen fractions (Fr.1–Fr.17) were collected every 2 min from 1 to 34 min with six replications. The fractions were dried with a steady flow of N_2 at room temperature. Each residue was then redissolved in 100 μL of 70% methanol followed by centrifugation at 14,000 rpm for 10 min, and the supernatant was retained for the second-dimensional (^{2}D)-RPLC separation.

The ^{2}D-RPLC separation was conducted on an UPLC Acquity™ system (Waters, USA) utilizing a Waters Acquity UPLC BEH C18 column (2.1 × 100mm, 1.7 μm). The mobile phase was composed of 0.1% formic acid (*v*/*v*) in water (A) and CH_3CN (B), with the column temperature at 40 °C. The elution gradient was as follows: 0–4 min, 90–80% A; 4–9 min, 80–77% A; 9–10 min, 77–71% A; 10–12 min, 71–70% A; 12–22 min, 70–68% A; 22–24 min, 68–66% A; 24–24.5 min, 66–58% A; 24.5–36.5 min, 58–15% A; 36.5–37.5 min, 15–5% A; 37.5–40.5 min, 5% A. Five μL were injected and the flow rate was 0.3 mL/min.

Acquisition was performed in MS^E mode with a Synapt™ QTOF high-resolution mass spectrometer (Waters, USA) under negative ion mode. The optimized parameters for mass detection were as follows: high-purity nitrogen (N_2) was used as desolvation gas (800 L/h) and nebulizer gas (40 psi); the desolvation temperature was 450 °C. The cone gas flow was set at 50 L/h, capillary voltage at 1.5 kV, cone voltage at 40 V, supplemental ion source voltage at 80V, ion source temperature at 120 °C, low collision energy at 6 eV, and high collision energy ranging from 50 to 70 eV. The mass scan range was *m*/*z* 350–1500. Real-time calibration was performed using leucine enkephalin (400 ng/mL) at a flow rate of 10 μL/min.

3.4. *Evaluation of Orthogonality and Peak Capacity*

Orthogonality and peak capacity of the developed 2D LC system were calculated with a set of asterisk formula [38,46] (Supplementary Formula (S1)). The normalized retention time (t_I) of each reference component to the relative retention time (t'_R, norm(i)) based on equation (Equation (S1)) (t_D: dead volume time; t_G: effective elution time of the chromatography system). The peak distribution around the four lines Sz_-, Sz_+, Sz_1, and Sz_2 were according to Equations (S2)–(S5) (σ: standard deviation of the values of all 23 index components). The Z parameters were calculated according to Equations (S6)–(S10), which yields the orthogonality result A_O. The peak capacity of theoretical ($n_{c,2D}$) and effective ($n'_{c,2D}$) were determined based on Equations (S11)–(S13), in which W_b represents the average peak widths of three well-separated chromatographic peaks at the beginning, middle, and end of the elution gradient.

3.5. Development of an In-House Database of P. genus

To comprehensively characterize triterpene saponins in RPJ, an in-house database of *P.* genus was established, including 612 saponins reported in the genus and 228 predicted metabolites (Tables S6 and S7). The database includes an Excel file with records of 840 saponins' names, formulas, theoretical molecular weights, and MS/MS characteristics, as well as a .mol file for each compound. The information on these saponins was primarily obtained through literature research. Chemical structures with incomplete information were obtained by retrieved from online databases such as PubChem (https://pubchem.ncbi.nlm.nih.gov/, accessed on 29 January 2024), ChemSpider (https://www.chemspider.com/, accessed on 29 January 2024) and Chemicalbook (https://www.chemicalbook.com/ProductIndex.aspx, accessed on 29 January 2024), or drawn using KingDraw 3.0 software. MS/MS fragments were acquired by searching online databases, including Massbank (https://massbank.jp, accessed on 6 March 2024), HMDB (https://hmdb.ca, accessed on 6 March 2024), etc.

Previous studies have demonstrated that OA-type saponins in the *P.* genus can undergo esterification with methyl, ethyl, and butyl groups [30], whereas dammarane-type ginsenosides may exhibit substitutions of malonyl and acetyl [22,36]. Consequently, potential structures of the saponins reported in PRJ were predicted, encompassing 47 OA-type saponin derivatives and 181 dammarane-type saponin derivatives. For predicted compounds, molecular weights were calculated using MassLynx 4.1 workstation and MS/MS fragments were inferred based on similar saponins.

3.6. Method Validation

Validation of the established HILIC × RP LC/QTOF-MS method was conducted for inter-/intraday precision, reproducibility and LOD using five reference saponins (G-Rh1, G-Re, CS-V, IVa and CS-IV) as index compounds. To evaluate the inter-/intraday precision of ^{1}D and ^{2}D separation, six repeated injections were performed on the first day, followed by three consecutive injections on the second and third days. Precision and reproducibility were assessed using the relative standard deviation (RSD, %). The LOD of the four reference saponins (G-Rh1, G-Re, CS-V, IVa and CS-IV) were determined at a signal-to-noise(S/N) ratio of about 3.

3.7. Automated Peak Annotation with UNIFI

MS^E data were recorded using Masslynx and then processed with UNIFI 1.8.2 software, which employed a three-dimensional peak apex track integration algorithm to detect the full-scan data and provide clear low and high-energy spectra [17]. UNIFI facilitated data correction, matching of precursor and product ions, and peak annotation based on an in-house database imported into the software. The parameters of automatic annotation were as follows: low-energy and high-energy ion intensity thresholds were set at 300 and 40 counts, respectively, and target match tolerance and fragment were set at 10.0 ppm. The adduct ions $[M - H]^-$ and $[M + HCOO]^-$ were used to automatically screen for target components. Following processing, the software generated a list of 'Identified Components'. To ensure accurate identification, a filter was applied with Detector Counts of $\geq$5000 and an error range of ppm $\leq$10.0 to reduce errors and false positives. The compounds listed under 'Unknown Components' were analyzed manually.

4. Conclusions

In this study, a sensitive and reliable offline HILIC × RP LC/QTOF-MS method was developed, along with an in-house database and structure prediction strategy. The method was successfully applied to characterize triterpene saponins from RPJ, demonstrating high orthogonality and peak capacity. A total of 307 saponins were identified from RPJ, with 76 of these saponins being identified for the first time in *P. japonicus*. These findings not only provide a deeper understanding of the chemical constituents of RPJ, but also offer a simple and effective approach for analyzing the complex composition of herbal medicine.

Supplementary Materials: The following supporting information can be downloaded at: https://www.mdpi.com/article/10.3390/molecules29061295/s1, Figure S1. ^{1}D HILIC chromatograms (203 nm) of RPJ extract with different columns: (a) Atlantis HILIC silica column (2.1 × 150 mm, 3 μm), (b) BEH Amide column (2.1 × 100 mm, 1.7 μm), and (c) Xbridge Amide column (2.1 × 150 mm, 3.5 μm); Figure S2. ^{1}D HILIC chromatograms (203 nm) of RPJ extract with different additives in the mobile phase: (a) Water; (b) 0.1% trifluoroacetic acid (TFA); (c) 0.1 M ammonium formate (AF); (d) 0.1% formic acid (FA); Figure S3. ^{1}D HILIC chromatograms (203 nm) of RPJ extract with XBridge Amide column at different column temperature: (a) 25 °C, (b) 30 °C, (c) 35 °C and (d) 40 °C; Figure S4. Base peak ion (BPI) chromatograms of 23 reference saponins (left) and RPJ extract (right) with different RP-UPLC columns at 40 °C. (a) Scepter C18-120 (2.1 × 150 mm, 1.9 μm); (b) CORTECS C18 column (2.1 × 100 mm, 1.6 μm); (c) BEH C18 column (2.1 × 100 mm, 1.7 μm); (d) BEH Shield RP18 column (2.1 × 100 mm, 1.7 μm); (e) HSS T3 column (2.1 × 100 mm, 1.8 μm); Figure S5. Base peak ion (BPI) chromatograms of 23 reference saponins (left) and RPJ (right) extract with BEH C18 column at different column temperature: (a) 30 °C, (b) 35 °C and (c) 40 °C; Formula (S1). Asterisk equations for calculating orthogonality and peak capacity of 2D LC system; Table S1. Intra-day precision of the ^{1}D separation (n = 6); Table S2. Inter-day precision of the ^{1}D separation (n = 3); Table S3. Intra-day precision of the ^{2}D separation (n = 6); Table S4. Inter-day precision of the ^{2}D separation (n = 3); Table S5. Repeatability of the offline 2D LC-MS system (n = 6); Table S6. Information of the 612 saponins reported from the *Panax* genus as of 2023; Table S7. 228 predicted saponin metabolites of *Panacis Japonici*; Table S8. Detailed structural information of the 307 saponins characterized by 2D LC-MS.

Author Contributions: Investigation and writing—original draft, S.Y. and C.L.; validation, S.W.; formal analysis, H.L. and Y.M.; data curation, Y.D. and J.C.; resources, P.Y.; project administration and writing—editing, H.Z. All authors have read and agreed to the published version of the manuscript.

Funding: The study was supported by the National Natural Science Foundation of China (Grant nos. 81973564 and 81403095) and the Scientific Research Program of Beijing Municipal Commission of Education (KM201910025020).

Institutional Review Board Statement: Not applicable.

Informed Consent Statement: Not applicable.

Data Availability Statement: The data presented in this study are available in article and Supplementary Materials.

Acknowledgments: We thank BioRender.com for the support of the diagram drawing.

Conflicts of Interest: The authors declare no conflicts of interest.

References

1. Tang, J.L.; Liu, B.Y.; Ma, K.W. Traditional Chinese medicine. *Lancet* **2008**, *372*, 1938–1940. [CrossRef] [PubMed]
2. You, L.; Liang, K.; An, R.; Wang, X. The path towards FDA approval: A challenging journey for Traditional Chinese Medicine. *Pharmacol. Res.* **2022**, *182*, 106314. [CrossRef] [PubMed]
3. Fu, Q.; Ke, Y.X.; Jiang, D.S.; Jin, Y. Chemical separation and characterization of complex samples with herbal medicine. *Trends Anal. Chem.* **2020**, *124*, 115775. [CrossRef]
4. Zuo, T.; Zhang, C.; Li, W.; Wang, H.; Hu, Y.; Yang, W.; Jia, L.; Wang, X.; Gao, X.; Guo, D. Offline two-dimensional liquid chromatography coupled with ion mobility-quadrupole time-of-flight mass spectrometry enabling four-dimensional separation and characterization of the multicomponents from white ginseng and red ginseng. *J. Pharm. Anal.* **2020**, *10*, 597–609. [CrossRef]
5. Pirok, B.W.J.; Stoll, D.R.; Schoenmakers, P.J. Recent Developments in Two-Dimensional Liquid Chromatography: Fundamental Improvements for Practical Applications. *Anal. Chem.* **2019**, *91*, 240–263. [CrossRef] [PubMed]
6. Montero, L.; Herrero, M. Two-dimensional liquid chromatography approaches in Foodomics—A review. *Anal. Chim. Acta* **2019**, *1083*, 1–18. [CrossRef] [PubMed]
7. Duarte, R.; Brandão, P.F.; Duarte, A.C. Multidimensional chromatography in environmental analysis: Comprehensive two-dimensional liquid versus gas chromatography. *J. Chromatogr. A* **2023**, *1706*, 464288. [CrossRef] [PubMed]

8. Qiu, S.; Yang, W.Z.; Shi, X.J.; Yao, C.L.; Yang, M.; Liu, X.; Jiang, B.H.; Wu, W.Y.; Guo, D.A. A green protocol for efficient discovery of novel natural compounds: Characterization of new ginsenosides from the stems and leaves of *Panax ginseng* as a case study. *Anal. Chim. Acta* **2015**, *893*, 65–76. [CrossRef] [PubMed]
9. Ji, S.; Wang, S.; Xu, H.S.; Su, Z.Y.; Tang, D.Q.; Qiao, X.; Ye, M. The application of on-line two-dimensional liquid chromatography (2DLC) in the chemical analysis of herbal medicines. *J. Pharm. Biomed. Anal.* **2018**, *160*, 301–313. [CrossRef] [PubMed]
10. Jia, L.; Wang, H.; Xu, X.; Wang, H.; Li, X.; Hu, Y.; Chen, B.; Liu, M.; Gao, X.; Li, H.; et al. An off-line three-dimensional liquid chromatography/Q-Orbitrap mass spectrometry approach enabling the discovery of 1561 potentially unknown ginsenosides from the flower buds of *Panax ginseng*, *Panax quinquefolius* and *Panax notoginseng*. *J. Chromatogr. A* **2022**, *1675*, 463177. [CrossRef] [PubMed]
11. Yang, Y.; Yang, L.; Zheng, M.; Cao, D.; Liu, G. Data acquisition methods for non-targeted screening in environmental analysis. *Trends Anal. Chem.* **2023**, *160*, 116966. [CrossRef]
12. Xu, Y.; Liu, Y.; Zhou, H.; Wang, R.; Yu, D.; Guo, Z.; Liang, X. A guide of column selection for two-dimensional liquid chromatography method development of natural alkaloids. *Talanta* **2023**, *251*, 123738. [CrossRef]
13. Chang, W.H.; Lee, C.Y.; Lin, C.Y.; Chen, W.Y.; Chen, M.C.; Tzou, W.S.; Chen, Y.R. UniQua: A universal signal processor for MS-based qualitative and quantitative proteomics applications. *Anal. Chem.* **2013**, *85*, 890–897. [CrossRef]
14. Guo, J.; Huan, T. Comparison of Full-Scan, Data-Dependent, and Data-Independent Acquisition Modes in Liquid Chromatography-Mass Spectrometry Based Untargeted Metabolomics. *Anal. Chem.* **2020**, *92*, 8072–8080. [CrossRef]
15. Zhang, C.; Zuo, T.; Wang, X.; Wang, H.; Hu, Y.; Li, Z.; Li, W.; Jia, L.; Qian, Y.; Yang, W.; et al. Integration of Data-Dependent Acquisition (DDA) and Data-Independent High-Definition MS(E) (HDMS(E)) for the Comprehensive Profiling and Characterization of Multicomponents from *Panax japonicus* by UHPLC/IM-QTOF-MS. *Molecules* **2019**, *24*, 2708. [CrossRef]
16. Wang, H.D.; Wang, H.M.; Wang, X.Y.; Xu, X.Y.; Hu, Y.; Li, X.; Shi, X.J.; Wang, S.M.; Liu, J.; Qian, Y.X.; et al. A novel hybrid scan approach enabling the ion-mobility separation and the alternate data-dependent and data-independent acquisitions (HDDIDDA): Its combination with off-line two-dimensional liquid chromatography for comprehensively characterizing the multicomponents from Compound Danshen Dripping Pill. *Anal. Chim. Acta* **2022**, *1193*, 339320. [PubMed]
17. López, M.G.; Fussell, R.J.; Stead, S.L.; Roberts, D.; McCullagh, M.; Rao, R. Evaluation and validation of an accurate mass screening method for the analysis of pesticides in fruits and vegetables using liquid chromatography-quadrupole-time of flight-mass spectrometry with automated detection. *J. Chromatogr. A* **2014**, *1373*, 40–50. [CrossRef] [PubMed]
18. Wu, Y.T.; Zhao, X.N.; Zhang, P.X.; Wang, C.F.; Li, J.; Wei, X.Y.; Shi, J.Q.; Dai, W.; Zhang, Q.; Liu, J.Q. Rapid Discovery of Substances with Anticancer Potential from Marine Fungi Based on a One Strain-Many Compounds Strategy and UPLC-QTOF-MS. *Mar. Drugs* **2023**, *21*, 646. [CrossRef] [PubMed]
19. Tsugawa, H.; Cajka, T.; Kind, T.; Ma, Y.; Higgins, B.; Ikeda, K.; Kanazawa, M.; VanderGheynst, J.; Fiehn, O.; Arita, M. MS-DIAL: Data-independent MS/MS deconvolution for comprehensive metabolome analysis. *Nat. Methods* **2015**, *12*, 523–526. [CrossRef] [PubMed]
20. Mahieu, N.G.; Genenbacher, J.L.; Patti, G.J. A roadmap for the XCMS family of software solutions in metabolomics. *Curr. Opin. Chem. Biol.* **2016**, *30*, 87–93. [CrossRef] [PubMed]
21. Röst, H.L.; Sachsenberg, T.; Aiche, S.; Bielow, C.; Weisser, H.; Aicheler, F.; Andreotti, S.; Ehrlich, H.C.; Gutenbrunner, P.; Kenar, E.; et al. OpenMS: A flexible open-source software platform for mass spectrometry data analysis. *Nat. Methods* **2016**, *13*, 741–748. [CrossRef] [PubMed]
22. Yao, C.L.; Pan, H.Q.; Wang, H.; Yao, S.; Yang, W.Z.; Hou, J.J.; Jin, Q.H.; Wu, W.Y.; Guo, D.A. Global profiling combined with predicted metabolites screening for discovery of natural compounds: Characterization of ginsenosides in the leaves of *Panax notoginseng* as a case study. *J. Chromatogr. A* **2018**, *1538*, 34–44. [CrossRef] [PubMed]
23. Committee, C.P. *Pharmacopoeia of the People's Republic of China*; The Medicine Science and Technology Press of China: Beijing, China, 2020; Volume 144.
24. Guo, Z.; Feng, Z.T.; Zhang, H.R.; Yan, L.; Liang, M.G.; Mei, Z.G.; Cai, S.J. Progress in the treatment of rheumatoid arthritis with Panax janonicus and its preparations. *J. Chin. Med. Mater.* **2019**, *42*, 941–944.
25. He, H.; Xu, J.; Xu, Y.; Zhang, C.; Wang, H.; He, Y.; Wang, T.; Yuan, D. Cardioprotective effects of saponins from *Panax japonicus* on acute myocardial ischemia against oxidative stress-triggered damage and cardiac cell death in rats. *J. Ethnopharmacol.* **2012**, *140*, 73–82. [CrossRef] [PubMed]
26. Wang, X.J.; Xie, Q.; Liu, Y.; Jiang, S.; Li, W.; Li, B.; Wang, W.; Liu, C.X. *Panax japonicus* and chikusetsusaponins: A review of diverse biological activities and pharmacology mechanism. *Chin. Herb. Med.* **2021**, *13*, 64–77. [CrossRef]
27. Wang, J.; He, L.Y.; Wang, S.Y.; Zhao, H.; Chen, J.; Dong, Y.X.; Yasen, S.; Wang, L.; Zou, H.Y. Therapeutic effect of the total saponin from *Panax japonicus* on experimental autoimmune encephalomyelitis by attenuating inflammation and regulating gut microbiota in mice. *J. Ethnopharmacol.* **2023**, *315*, 116681. [CrossRef] [PubMed]
28. Wang, L.; Zhao, H.; Zou, H.Y.; Zhang, Q.; Zheng, Q.; Li, Q.; Fang, L. *New Use of Extract of Panax japonicus as Multiple Sclerosis Drug*; China National Intellectual Property Administration: Beijing, China, 2016.
29. Li, X.; Liu, J.; Zuo, T.T.; Hu, Y.; Li, Z.; Wang, H.D.; Xu, X.Y.; Yang, W.Z.; Guo, D.A. Advances and challenges in ginseng research from 2011 to 2020: The phytochemistry, quality control, metabolism, and biosynthesis. *Nat. Prod. Rep.* **2022**, *39*, 875–909. [CrossRef] [PubMed]

30. He, Y.M.; Ai, K.; He, C.X.; Zhang, C.C.; Yuan, D.; Cai, S.J. Triterpene saponins in *Panax japonicus* and their ^{13}C-NMR spectroscopic characteristics. *China J. Chin. Mater. Medica* **2019**, *44*, 249–260.
31. Chen, J.L.; Tan, M.X.; Zou, L.S.; Liu, X.H.; Chen, S.Y.; Shi, J.J.; Wang, C.C.; Mei, Y.Q. Saponins in *Panacis japonici* Rhizoma as Analyzed by UFLC-Triple TOF MS/MS. *Food Sci.* **2019**, *40*, 249–258.
32. Du, Z.; Li, J.; Zhang, X.; Pei, J.; Huang, L. An Integrated LC-MS-Based Strategy for the Quality Assessment and Discrimination of Three *Panax* Species. *Molecules* **2018**, *23*, 2988. [CrossRef] [PubMed]
33. Yang, W.; Zhang, J.; Yao, C.; Qiu, S.; Chen, M.; Pan, H.; Shi, X.; Wu, W.; Guo, D. Method development and application of offline two-dimensional liquid chromatography/quadrupole time-of-flight mass spectrometry-fast data directed analysis for comprehensive characterization of the saponins from Xueshuantong Injection. *J. Pharm. Biomed. Anal.* **2016**, *128*, 322–332. [CrossRef] [PubMed]
34. Xie, G.; Plumb, R.; Su, M.; Xu, Z.; Zhao, A.; Qiu, M.; Long, X.; Liu, Z.; Jia, W. Ultra-performance LC/TOF MS analysis of medicinal *Panax* herbs for metabolomic research. *J. Sep. Sci.* **2008**, *31*, 1015–1026. [CrossRef] [PubMed]
35. Falev, D.I.; Voronov, I.S.; Onuchina, A.A.; Faleva, A.V.; Ul'yanovskii, N.V.; Kosyakov, D.S. Analysis of Softwood Lignans by Comprehensive Two-Dimensional Liquid Chromatography. *Molecules* **2023**, *28*, 8114. [CrossRef] [PubMed]
36. Yang, W.Z.; Bo, T.; Ji, S.; Qiao, X.; Guo, D.A.; Ye, M. Rapid chemical profiling of saponins in the flower buds of *Panax notoginseng* by integrating MCI gel column chromatography and liquid chromatography/mass spectrometry analysis. *Food Chem.* **2013**, *139*, 762–769. [CrossRef] [PubMed]
37. Xu, J.D.; Xu, M.Z.; Zhou, S.S.; Kong, M.; Shen, H.; Mao, Q.; Zhu, H.; Chan, G.; Liu, L.F.; Zhang, Q.W.; et al. Effects of chromatographic conditions and mass spectrometric parameters on the ionization and fragmentation of triterpene saponins of *Ilex asprella* in liquid chromatography-mass spectrometry analysis. *J. Chromatogr. A* **2019**, *1608*, 460418. [CrossRef] [PubMed]
38. Camenzuli, M.; Schoenmakers, P.J. A new measure of orthogonality for multi-dimensional chromatography. *Anal. Chim. Acta* **2014**, *838*, 93–101. [CrossRef] [PubMed]
39. Wang, M.; Xu, X.Y.; Wang, H.D.; Wang, H.M.; Liu, M.Y.; Hu, W.D.; Chen, B.X.; Jiang, M.T.; Qi, J.; Li, X.H.; et al. A multi-dimensional liquid chromatography/high-resolution mass spectrometry approach combined with computational data processing for the comprehensive characterization of the multicomponents from *Cuscuta chinensis*. *J. Chromatogr. A* **2022**, *1675*, 463162. [CrossRef] [PubMed]
40. Li, Y.J.; Wei, H.L.; Qi, L.W.; Chen, J.; Ren, M.T.; Li, P. Characterization and identification of saponins in *Achyranthes bidentata* by rapid-resolution liquid chromatography with electrospray ionization quadrupole time-of-flight tandem mass spectrometry. *Rapid. Commun. Mass. Spectrom.* **2010**, *24*, 2975–2985. [CrossRef]
41. Zou, H.Y.; Zhao, H.; Qiu, K.; Cui, X.G.; Yu, P. Study on Saponins of *Panax japonicus* Rhizome. *World Chin. Med.* **2012**, *7*, 565–566.
42. Liu, Y.; Song, Q.; Liu, W.; Li, P.; Li, J.; Zhao, Y.; Zhang, L.; Tu, P.; Wang, Y.; Song, Y. Authentic compound-free strategy for simultaneous determination of primary coumarins in Peucedani Radix using offline high performance liquid chromatography-nuclear magnetic resonance spectroscopy-tandem mass spectrometry. *Acta. Pharm. Sin. B* **2018**, *8*, 645–654. [CrossRef] [PubMed]
43. Yang, H.; Lee, D.Y.; Kang, K.B.; Kim, J.Y.; Kim, S.O.; Yoo, Y.H.; Sung, S.H. Identification of ginsenoside markers from dry purified extract of *Panax ginseng* by a dereplication approach and UPLC-QTOF/MS analysis. *J. Pharm. Biomed. Anal.* **2015**, *109*, 91–104. [CrossRef] [PubMed]
44. Yoshizaki, K.; Yahara, S. New triterpenoid saponins from fruits specimens of *Panax japonicus* collected in Kumamoto and Miyazaki prefectures (1). *Chem. Pharm. Bull.* **2012**, *60*, 354–362. [CrossRef] [PubMed]
45. Nakamura, S.; Sugimoto, S.; Matsuda, H.; Yoshikawa, M. Medicinal flowers. XVII. New dammarane-type triterpene glycosides from flower buds of American ginseng, *Panax quinquefolium* L. *Chem. Pharm. Bull.* **2007**, *55*, 1342–1348. [CrossRef] [PubMed]
46. Li, X.; Stoll, D.R.; Carr, P.W. Equation for peak capacity estimation in two-dimensional liquid chromatography. *Anal. Chem.* **2009**, *81*, 845–850. [CrossRef] [PubMed]

molecules

MDPI

Article

The Validation and Determination of Empagliflozin Concentration in the Presence of Grapefruit Juice Using HPLC for Pharmacokinetic Applications

Wael Abu Dayyih [1,*,†], Zainab Zakaraya [2], Mohammad Hailat [3,†], Nafe M. Al-Tawarah [4], Sahem Alkharabsheh [4], Haya Khalid Nadher [5], Zeyad Hailat [6], Samia M. Alarman [1], Anas Khaleel [5] and Riad Awad [5]

1 Faculty of Pharmacy, Mutah University, Al-Karak 61710, Jordan; erman.samia@yahoo.com
2 Faculty of Pharmacy, Al-Ahliyya Amman University, Amman 19328, Jordan; z.zakaraya@ammanu.edu.jo
3 Faculty of Pharmacy, Al-Zaytoonah University of Jordan, Amman 11733, Jordan; m.hailat@zuj.edu.jo
4 Faculty of Allied Medical Sciences, Medical Laboratory Sciences, Mutah University, Al-Karak 61710, Jordan; nafitawa77@gmail.com (N.M.A.-T.); sahem@mutah.edu.jo (S.A.)
5 Faculty of Pharmacy and Medical Sciences, University of Petra, Amman 11196, Jordan; anas.khaleel@uop.edu.jo (A.K.); rawad@uop.edu.jo (R.A.)
6 Department of Information Systems, Yarmouk University, Irbid 21163, Jordan; zeyad.hailat@yu.edu.jo
* Correspondence: wabudayyih@mutah.edu.jo
† These authors contributed equally to this work.

Abstract: Type 2 diabetes mellitus is a multifactorial disorder whose primary manifestation usually initiates with elevated blood sugar levels. Several antidiabetic agents are used to manage type 2 diabetes mellitus, of which empagliflozin is an oral sodium-glucose co-transporter (SGLT-2) inhibitor in the kidney. This research aims to develop and validate a simple analytical method for determining empagliflozin levels in biological fluid and to further evaluate grapefruit juice's impact on empagliflozin pharmacokinetics in rats. High-Performance Liquid Chromatography (HPLC) was used to establish a simple, rapid, and accurate method for determining empagliflozin levels in rat plasma, in the presence of grapefruit juice. Four groups of rats (n = 10 rats in each) were used in the preclinical study. Group A (healthy rats) received empagliflozin alone; Group B (healthy rats) received empagliflozin with grapefruit; Group C (diabetic rats) received empagliflozin with grapefruit; and Group D (healthy, negative control) received no medication. The rats (n = 10) were given grapefruit juice instead of water for seven days before receiving the empagliflozin dose (0.16 mg/kg). Some pharmacokinetic parameters for each group were determined. The maximum plasma concentration (C_{max}) and area under the curve (AUC) of empagliflozin in Group A without grapefruit intake were 730 ng/mL and 9264.6 ng × h/mL, respectively, with T_{max} (2 h). In Group B, C_{max} was 1907 ng/mL and AUC was 10,290.75 ng × h/mL in the presence of grapefruit, with T_{max} (1 h); whereas, in Group C, the C_{max} was 2936 ng/mL and AUC was 18657 ng × h/mL, with T_{max} (2 h). In conclusion, our results showed that the co-administration of grapefruit with empagliflozin should be cautiously monitored and avoided, in which grapefruit elevates the plasma level of empagliflozin. This may be attributed to the inhibition of the uridine enzyme in the grapefruit by hesperidin, naringin, and flavonoid.

Keywords: type 2 diabetes mellitus; sodium-glucose co-transporter inhibitor; (SGLT-2) inhibitor; empagliflozin; pharmacokinetics

1. Introduction

Diabetes mellitus (DM) is one of the most common chronic disorders, usually associated with elevated blood sugar concentrations [1]. DM is mainly caused by insufficient insulin production from the pancreas and low cell sensitivity to the naturally secreted insulin [2].

The inhibition of the sodium-glucose co-transporter (SGLT2) permits an increased excretion of renal glucose, leading to lowered blood glucose levels. SGLT2 regulates most renal glucose regeneration. The blood glucose decreases renal re-absorption and stimulates the kidney's carrier protein, resulting in Urinary Glucose Exclusion [3]. The potency of its management is independent of insulin secretion and operation. This mechanism allows 1,3-biphosphoglycerate (BPG) to be combined with other antidiabetic therapies, giving the best management a complementary benefit. The SGLT2i function without insulin secretion is unaffected by β-cell depletion and insulin signaling desensitization [4].

Empagliflozin was the first antidiabetic drug to minimize cardiovascular and overall mortality in T2DM patients [5] with elevated cardiovascular risk, confirmed by new and significant clinical trials with SGLT2i in preventing hyperglycemia-induced risks [6]. No therapy has demonstrated comparable reductions in cardiovascular and overall mortality in T2DM patients with proven cardiovascular risk to date in either a dipeptidyl peptidase-4 (DPP-4) inhibitor or a glucagon-like peptide-1 (GLP-1) [7].

Pharmacokinetic or pharmacodynamic mechanisms can mediate drug–food interactions. A drug's absorption difference may be clinically significant when considering its dosage and interactions with food [7]. For instance, slow-release theophylline formulations may differ in efficacy and can change upon consumption; the effectiveness of cyclosporine medicines may vary significantly based on the form taken and what is being co-administered with it. On the other hand, the effects of the medication on blood pressure or blood sugar levels in most patients are rarely clinically meaningful, as long as a rapid onset of those effects is not required [8].

The pharmacodynamics of certain drugs depend on the pre-existing chemistry of the body to generate the desired effect. When there is an incompatibility between a drug and the food taken by the patient, a pharmacodynamic interaction may occur [9], causing an antagonistic impact [8].

Due to the significant increase in grapefruit harvesting in the last decade, it has gained tremendous popularity [10]. Grapefruit enhances the bioavailability of some medications that interact with cytochromes (CYP450-3A4) [11]. The biological interest in grapefruit has encouraged chemical discovery, separation, and the characterization of several new substances [12].

This work aims to develop and validate a simple analytical method for determining empagliflozin levels in biological fluid using HPLC and also to evaluate grapefruit juice's impact on empagliflozin pharmacokinetics in rats. We used a bio-analytical method for studying the effect of grapefruit on the pharmacokinetic parameters of empagliflozin, an SGLT2 inhibitor in type 2 diabetes mellitus in rats.

2. Results

2.1. Results of Validation

A partial method validation was performed to demonstrate the reliability of the stated HPLC method for determining empagliflozin and grapefruit concentrations in rat plasma using the parameters indicated in Table 1.

2.2. Accuracy and Precision

The method's precision and accuracy were calculated by analyzing six samples with three replicates, each by two people on two days. The standard deviation (SD)-to-mean ratios were used to measure the relative standard deviation values (RSD) or CV %, expressed as percentages. For concentration and accuracy, the appropriate CV % limits, which should be less than 1.5, were determined. Furthermore, the agreed accuracy criterion of 85–115% for all concentrations was met.

The Quality Control Law (QCL) for empagliflozin analysis across six samples was performed. The analysis includes parameters such as the sample area, empagliflozin area, internal standard (IS) area, area ratio, actual concentration in nanograms per milliliter (ng/mL), theoretical concentration, accuracy, average accuracy, and relative standard deviation (RSD).

Table 1. Chromatographic condition summary.

Mobile phase composition	1 mL of triethylamine adjusted to pH 3.5 using orthophosphate, then (50:50 *v*/*v*) acetonitrile–potassium dihydrophosphate buffer.
Column type	C18 column (Hypersil-Silica, C-18, 250 mm × 4.6 mm, particle size—5 µm)
Wavelength	230 nm
Pump flow rate	0.75 mL/min
Auto-sampler temperature	25 °C
Column oven temperature	25 °C
Auto-sampler injection volume	20 µL
Retention Times (min)	
Metformin	2.4 min
Empagliflozin	3.7 min

2.3. *Absolute Recovery (Result of Matrix Effect)*

The absolute recovery was calculated by measuring the fundamental empagliflozin peak region and an internal standard using an analytical method based on the plasma samples prepared for a drug concentration or an internal standard to ensure 100% recovery in the peak areas with pure standards. The degree to which the empagliflozin and the internal standard are recovered should be consistent, exact, and replicable. The accuracy measured at each level should not exceed 15% of the variance coefficient (percentage of CV). Absolute recovery (the result of the matrix effect) is illustrated in Table 2 below.

Table 2. Absolute recovery (result of matrix effect, $n = 7$).

	IS-Normalized Empagliflozin		
	Mean	SD	RSD%
QC_{Low}	270.14	10.19	3.8
QC_{High}	865	8.64	1.0
Absolute recovery	101.9	3.6	3.5
	100.6	0.9	0.9

2.4. *Grapefruit Effect on Empagliflozin Pharmacokinetics*

Grapefruit juice is one of the most thoroughly researched dietary substances inhibiting CYP3A4 enteric metabolism.

2.5. *Group A (Empagliflozin Alone)*

According to the results shown in Figure 1, the maximum concentration of empagliflozin C_{max} 730 ng/mL was reached two hours after administration. A total of 26 h later, it gradually reached 49.2 ng/mL, the minimum concentration of empagliflozin. The area under the curve after 96 h ($AUC_{0–96}$) was found to be 9264.6 ng × h/mL, as shown in Figure 1.

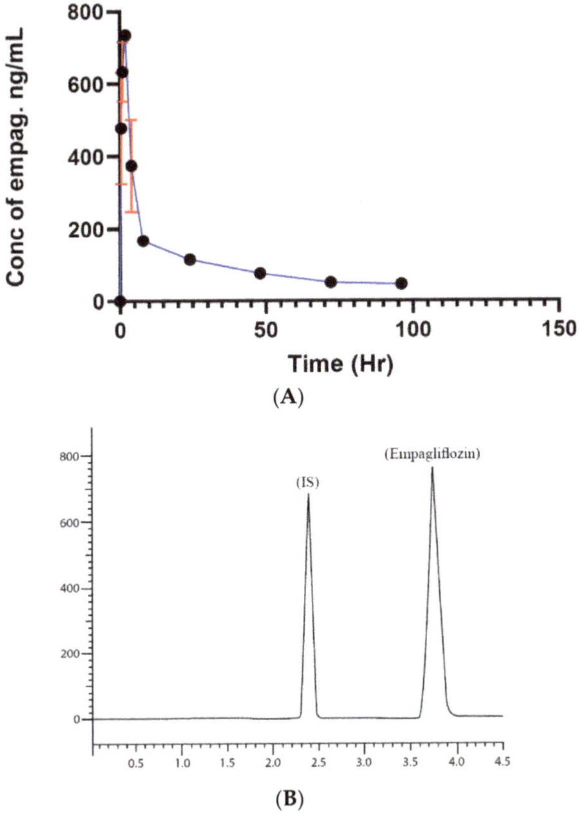

Figure 1. (**A**) Empagliflozin concentration vs. time plot. (n = 10), data ± SD, (**B**) HPLC chromatogram shows peaks and retention times for metformin (internal standard, IS) and empagliflozin.

2.6. Group B (Empagliflozin and Grapefruit) in Normal Rats

According to the results shown below in Figure 2, the maximum concentration of empagliflozin C_{max} 1907 ng/mL was reached one hour after administration. A total of 26 h later, it gradually reached 45 ng/mL, recorded as the minimum concentration of empagliflozin. The area under the curve after 96 h ($AUC_{0–96}$) was found to be 10,290.75 ng × h/mL, as shown in Figure 2.

2.7. Group C (Empagliflozin and Grapefruit) on Diabetes-Induced Rats

According to the results below in Figure 3, the maximum concentration of empagliflozin C_{max} 2936 ng/mL was reached two hours after administration. A total of 96 h later, it reached 47 ng/mL gradually, accordingly recorded as the minimum concentration of empagliflozin to be achieved. The area under the curve after 96 h ($AUC_{0–96}$) was 18,657 ng × h/mL, as shown in Figure 3.

Comparing empagliflozin pharmacokinetic parameters alone in normal rats in Group A and empagliflozin pharmacokinetic parameters in the presence of grapefruit on normal rats in Group B, the drug plasma level was increased in the presence of grapefruit. The T_{max} in Group B decreased to one hour and the C_{max} in Group B (1907 ng/mL) was increased due to the effect of grapefruit on the drug plasma level. The AUC for Group B (10,290.75 ng × h/mL) was also increased due to the impact of the grapefruit.

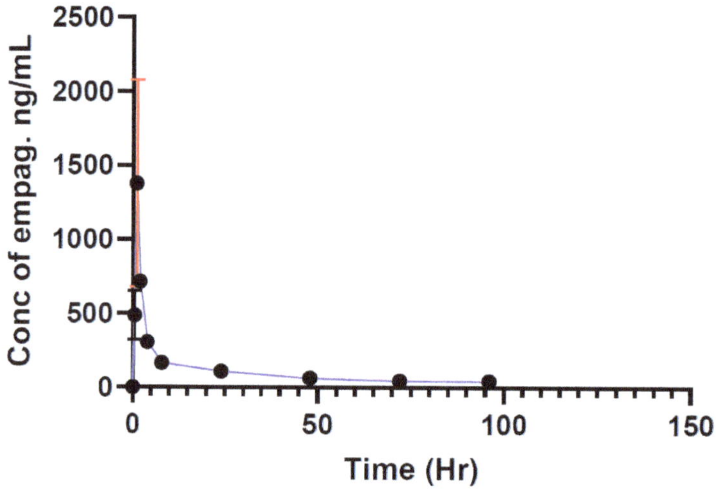

Figure 2. Empagliflozin concentration for normal rats treated with grapefruit vs. time (Group B, $n = 10$). Data ± SD.

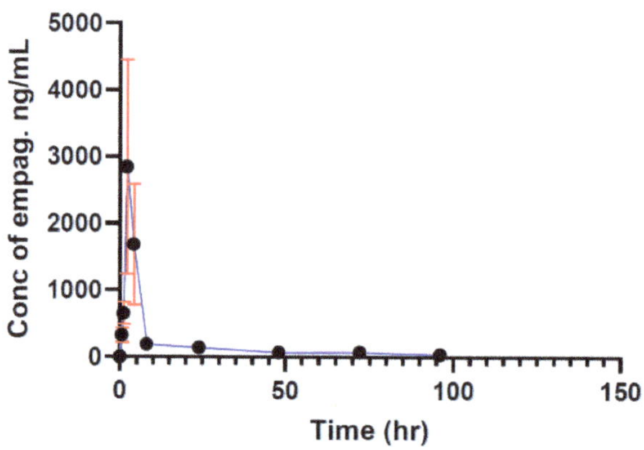

Figure 3. Empagliflozin concentration (grapefruit effect in diabetic rats, Group C) vs. time. $n = 10$, data ± SD.

Further comparing empagliflozin pharmacokinetic parameters with grapefruit on normal rats in Group B and with grapefruit in diabetes-induced rats in Group C, C_{max} was increased in induced rats in Group C. The T_{max} in Group C was increased in comparison to Group B, to become 2 h, the C_{max} and AUC in Group C were equal to 2936 ng/mL and 18,657 ng × h/mL, indicating that there is a highly significant increase in the C_{max} plasma and AUC as compared statistically, and as shown in Figures 4 and 5.

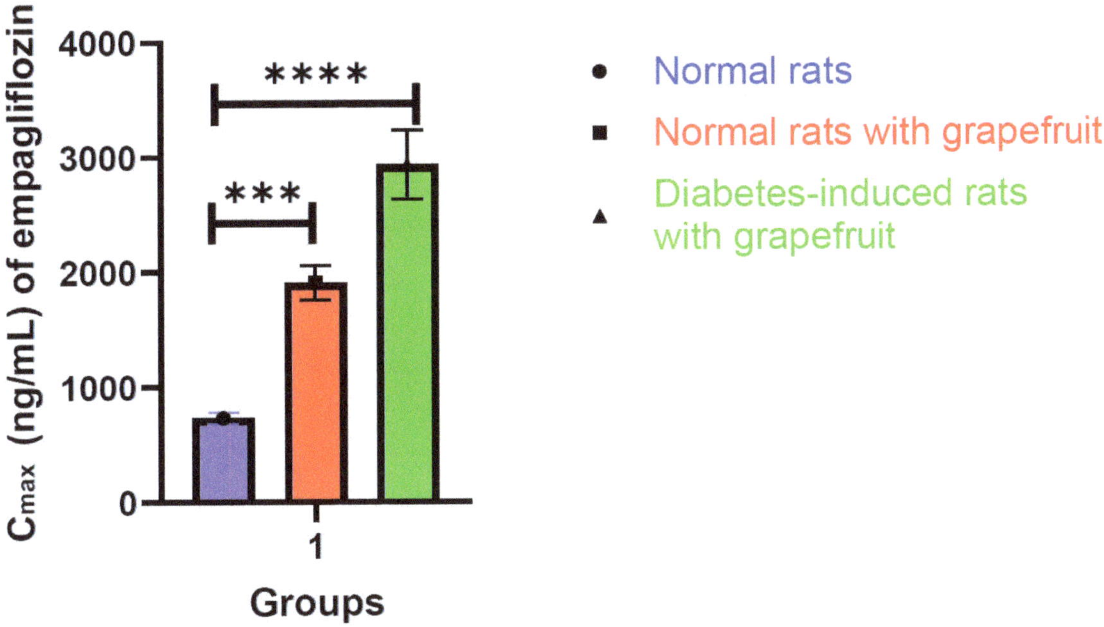

Figure 4. Histogram showing the average C_{max} of empagliflozin in normal rats, normal rats given grapefruit, and diabetes-induced rats given grapefruit (mean ± SD; ***, ****, indicates the *p*-value is less than or equal to 0.001, 0.0001, respectively).

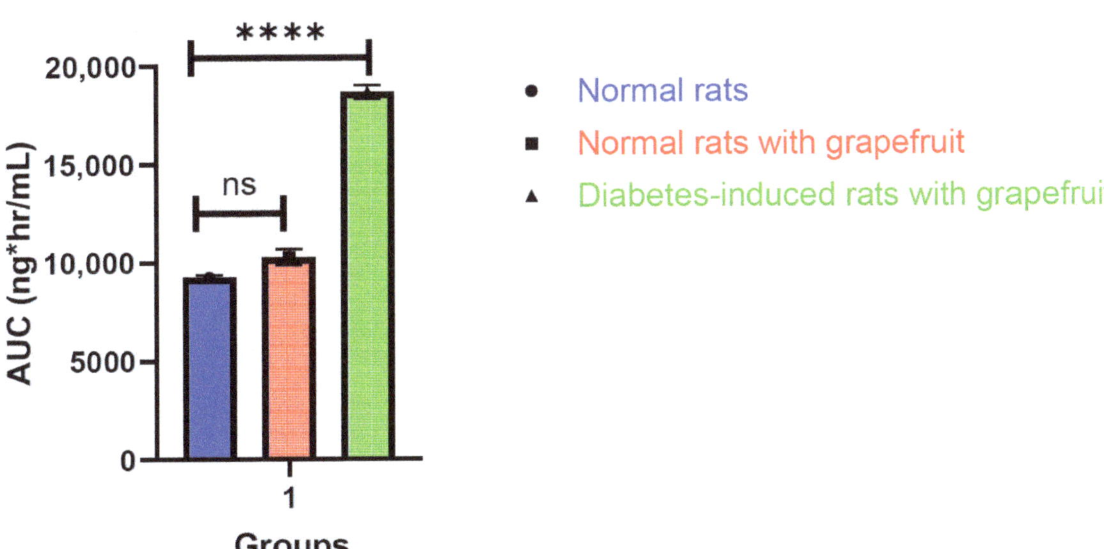

Figure 5. Histogram showing the AUC of empagliflozin in normal rats, normal rats given grapefruit, and diabetes-induced rats given grapefruit (mean ± SD; ****, indicates the *p*-value is less than or equal to 0.0001, respectively. ns: not significant).

3. Discussion

A partial, rapid, simple, and accurate method for determining empagliflozin levels in rat plasma has been developed using a High-Performance Liquid Chromatography–UV detector. Grapefruit administration affected the plasma empagliflozin level. When grapefruit was added to empagliflozin, the plasma Sitagliptin level increased dramatically compared to empagliflozin alone, especially in the DM-induced group.

Notably, the measured concentrations in the Quality Control Law (QCL) for empagliflozin analysis exhibit consistency with the theoretical values, indicating the robustness of the analytical method. The average accuracy of 103% reflects the precision and reliability of the analysis, while the low RSD of 0.30% highlights the minimal variability between measurements. These findings suggest a high level of confidence in the accuracy and precision of the empagliflozin analysis using the Quality Control Law approach.

The induced DM rats' group, which takes empagliflozin with grapefruit, exhibited significant differences in the C_{max} and AUC compared to both normal groups, regardless of whether empagliflozin was administered with or without grapefruit. The C_{max} also increased in the presence of grapefruit in the normal group, but less than in the DM-induced group. From the results above, we can conclude that the plasma level of empagliflozin increases in the case of grapefruit intake in the induced group rather than the normal group because of the UGT2B7, UGT1A3, UGT1A8, and UGT1A9 enzyme inhibition, which may be associated with a change in the absorption profile of the drug, especially in the diabetic rats.

Empagliflozin is a substrate metabolized by uridine 50-diphosphate-glucuronosy ltransferases that may affect empagliflozin's pharmacokinetics and pharmacodynamics [13,14]. It is also a substrate for p-gp, BCRP, OATP1B3, and 1B1 in the intestine, partially responsible for empagliflozin's active absorption [15,16]. Grapefruit juice is reported to have an inhibitory effect on p-gp efflux [17,18], a mechanism that might be involved in the absorption of this drug; this can partly explain the increase in the C_{max}, accompanied by an increase in the rate of absorption, expressed as a shortening of the T_{max} [19]. The AUC could also be somewhat increased due to this reason as it increased slightly but significantly ($p < 0.05$).

As described earlier, empagliflozin metabolism is mediated through UGT isomers in the liver, where empagliflozin is also reported to have an inhibitory effect on them [20,21]. This explains the increase in the AUC and Kel in both groups treated with empagliflozin and grapefruit. Enzyme inhibition may result from the interaction of hesperidin and naringenin, which would result in a higher amount of the drug in the body for a longer time, explained by an increase in the AUC and a decrease in the Kel, as well as an increase in elimination half-lives [22].

This effect was of a higher magnitude in diabetic rats, possibly due to the changes in DM-related enzyme levels. The T_{max} of Group C was longer than that of Group B, possibly due to the exact reasons for the changed absorption pattern via transporters and the high effect of enzyme inhibition, which gave a greater extent of bioavailability, with a very long half-life. Our study has some limitations; for instance, only one model was used, and a more advanced model should be used in future studies to confirm the findings further.

When these results are compared to previously described results elsewhere [11] for sitagliptin (a competitive inhibitor of the dipeptidyl peptidase 4 (DPP-4) enzyme), several differences and similarities emerge, as shown in Table 3.

Both studies (this article and a previous one [11]) discovered that co-administering the medication with grapefruit juice increased drug levels (the C_{max} and AUC). However, the degree of increase differed between studies. The empagliflozin research revealed a more pronounced effect, particularly in the diabetic group. The sitagliptin study showed a statistically significant increase in the Cmax and AUC. However, the T_{max} decreased by half in the healthy group in the empagliflozin study, whereas the T_{max} did not shows a statistically significant increase in the sitagliptin study. Both studies indicate a clinically significant interaction, emphasizing the need to avoid grapefruit juice while taking these medications.

Table 3. Comparisons between the effect of grapefruit juice co-administration on empagliflozin and sitagliptin pharmacokinetics [11].

Feature	Empagliflozin Study	Sitagliptin Study
Drug Dose	0.16 mg/kg	5.75 mg/kg
Rat Groups	Healthy (alone), Healthy (grapefruit), Diabetic (grapefruit), Negative Control	Control, Grapefruit Juice
Grapefruit Juice Administration	Seven days pre-dose	Grapefruit was given to the B group instead of drinking water two days before the study
Impact of Grapefruit Juice on the C_{max}, AUC, and T_{max} compared with the controls	C_{max} increased, AUC increased, T_{max} decreased	C_{max} increased, AUC increased, T_{max} not changed
Grapefruit Juice Impact on the C_{max}	Doubled in both healthy and diabetic groups	Significantly increased
Grapefruit Juice Impact on the AUC	Doubled in the healthy group, tripled in the diabetic group	Significantly increased
Grapefruit Juice Impact on the T_{max}	Decreased by half in the healthy group	No significant increase
Conclusion	Avoid co-administration due to significant increase in drug levels	Drug–food interaction observed; avoid grapefruit juice at the same time.

To summarize, the data indicate that giving empagliflozin with grapefruit juice increased drug levels (the C_{max} and AUC). However, the degree of increase differed amongst the treatment groups. The research indicated a more dramatic effect, particularly in the diabetic group. T_{max} was lowered by half in the healthy group. The study discovered a clinically significant interaction and suggested avoiding grapefruit juice when taking these medications. The summary of the research results is shown in Table 4.

Table 4. Pharmacokinetics summary of the four rat groups.

Group Number and Description	Pharmacokinetics Parameter Summary for Empagliflozin after Administration of Specific Treatment(s)
Group (A): healthy—treated with empagliflozin 0.5 mL (0.16 mg/mL) only	C_{max} (730 ng/mL), AUC (9264.6 ng × h/mL), T_{max} (2 h)
Group (B): healthy—treated with grapefruit juice (10 mL/day) for four days. On the fourth day, they were treated with empagliflozin 0.5 mL (0.16 mg/mL)	C_{max} (1907 ng/mL), AUC (10,290.75 ng × h/mL), T_{max} (1 h)
Group (C): A diabetic group was treated with grapefruit juice (10 mL/day) for four days (grapefruit juice replaced water for seven days). On the fourth day, the group was treated with 0.5 mL of empagliflozin (25 mg/150 mL).	C_{max} (2936 ng/mL), AUC (18,657 ng × h/mL), T_{max} (2 h)
Group (D): healthy (negative control) no drugs were given	no drugs were given

More research may be needed to investigate the interaction mechanisms, individual patient variability in response, and the concentration of enzymes related to the metabolism and absorption of the empagliflozin drug.

4. Materials and Methods

Empagliflozin purity is 99.5% and was purchased from Sigma-Aldrich® (Steinheim, Germany). Metformin HCL (99.5%) was obtained as a gift from Dar Al Dawa Pharmaceutical Company (Amman, Jordan). HPLC grade triethyl amine and HPLC grade methanol were from Fisher Scientific (Leicestershire, UK). Acetonitrile and Orthophosphate acid were purchased from Fisher Scientific (Leicestershire, UK). VWR International (Lutterworth, Leicestershire, UK) provided water (HiPerSolv CHROMANORM for HPLC) as the HPLC solvent. Streptozotocin (>95%; (bioXTra, London, UK), Lot # 18883-66-4) were purchased and used. All of the other chemicals were of reagent grade and used as received.

4.1. Instruments

An HPLC (FINNIGAN SURVEYOR) Liquid Chromatograph (Thermo Electron Corporation, San Jose, CA, USA) consists of a reciprocating quaternary gradient pump (LC Pump Plus) (Solvent delivery system pump), an auto-sampler (Auto-sampler Plus), a thermostatically controlled oven, a detector (UV-VIS Plus Detector), a communication bus module (CBM-20A), and a C18 column (Hypersil-Silica, C-18, 250 mm × 4.6 mm, particle size—5 µm, Thermo-Fisher, Cleveland, OH, USA). In addition, a Single-Pan Digital Balance (Sartorius) and an ultra-violet (UV) spectrophotometer (V530, version. 1.50.00, JASCO, Tokyo, Japan) controlled by Windows NT-based spectra manager were also used. A pH meter (model Sartorius 7110) was also used to measure pH. Centrifuge (M-24A, Boeco, Hamburg, Germany). Vortexes (Labinco, Breda, The Netherlands) and a sonicator (Elmasonic S100, Patterson, NJ, USA) were also used. The analysis was conducted at the University of Petra Pharmaceutical Centre's Instrumental Laboratory.

Glassware such as volumetric flasks, funnels, beakers (Isolab, Wertheim, Germany Class A, DIN), pipettes (Isolab, Wertheim, Germany Class 2Aa, DIN), and micro-pipettes (Socoerx, ISBA S.A., Ecubleus, Switzerland) of 100 µL and 1000 µL capacity were used.

4.2. Animal Handling

The animals are described in detail in Table 5. The protocol of the study was approved by the ethical committee of the Scientific Research and Ethics Committee [SREC]–Faculty of Pharmacy/Mutah University (NO. SREC1132023, date 13 April 2023). The preclinical study was conducted at the Animal House of the Applied Science University. All the experiments were conducted as per the University of Petra and Applied Science Private University institutional guidelines on animal use, which adopt the Federation of European Laboratory Animal Science Association (FELASA) guidelines.

Table 5. Animal description.

Animal Species	Wisteria Rat
Weight	200 g
Number of animals	40
Gender	Males
Age	Eight weeks

The rats were divided into four groups after marking each on its tail for identification; each group contained seven rats.

Group A: healthy—treated with empagliflozin 0.5 mL (0.16 mg/mL).

Group B: healthy—treated with grapefruit juice (10 mL/day) for four days. On the fourth day, they were treated with empagliflozin 0.5 mL (0.16 mg/mL).

Group C: Diabetic-induced group was treated with Streptozocin for three days at a dose of 45 mg; Streptozocin solution was prepared by dissolving 45 mg of Streptozocin in 12 mL of sodium citrate buffer and was mixed in a blender to obtain a homogenous solution of 3.75% (*w*/*v*), the concentration of Streptozocin was 3.75 mg/1 mL. As concerns the solution injected IP for each rat, after DM induction, the group was treated with grapefruit juice (10 mL/day) for four days (grapefruit juice will be given instead of water for seven days). On the fourth day, the group was treated with empagliflozin 0.5 mL (25 mg/150 mL). The Accu-Chek® device was used for blood glucose measurements.

Group D: healthy—control no drug given (negative control).

4.3. Collection of Blood Samples

Blood samples were taken from the rats' tails on the first day at defined time intervals, as follows: 0.5 h, 1 h, 2 h, 4 h, 6 h, 8 h, 24 h, 48 h, 72 h, and 96 h.

The samples were collected into an EDTA tube and centrifuged immediately at a speed of 5000 rounds per minute (RPM) for 10 min.

Plasma was obtained, placed into a labeled Eppendorf tube, and stored at −30 °C until analysis.

4.4. Sample Preparation

4.4.1. Mobile Phase Preparation

Depending on previous studies [23] and trying several times to prepare the mobile phase, we found that the optimum buffer for the mixture of the prepared mobile phase was potassium dihydrophosphate buffer, which was designed by dissolving potassium dihydrophosphate salt (KH_2PO_4) in 1 L distilled water, then 1 mL of triethylamine added and, afterwards, the mixture was pH adjusted to 3.5, using orthophosphate. The mobile phase of acetonitrile–potassium dihydrophosphate buffer (50:50 v/v) was prepared. The flow rate was set to 0.75 mL/min, and the detection was achieved using a UV detector at 230 nm, with metformin as an internal standard. The time taken for the completion of the analysis was below five minutes. Metformin HCL and empagliflozin were identified using UV spectrum, peak purity, and retention times. All these chromatographic conditions were performed at ambient room temperature.

4.4.2. Selection of Wavelength (λ) for the Chromatography

From the UV spectrum recorded on the UV–visible spectrophotometer, 230 nm was a suitable wavelength for detection. A solution containing 50 μg/mL of metformin HCL and empagliflozin in aqueous methanol (50%) was injected into the HPLC system and peak parameters were monitored at 230 nm.

4.4.3. Preparation of Drug Solution

The rats were given 0.5 mL daily of 0.166 mg/mL empagliflozin, which means that we prepared a solution containing 25 mg of empagliflozin per 150 mL by milling an empagliflozin tablet (Jardiance® 25 mg) and subsequently dissolving it in 150 mL of distilled water.

4.4.4. Stock Working Solution Preparation of Empagliflozin

The empagliflozin stock solution was prepared by dissolving 100 mg of empagliflozin in 100 mL of methanol. The resulting solution contained 1×10^6 ng/mL of empagliflozin.

4.4.5. Stock Working Solution Preparation of Metformin

The metformin (IS) stock solution was prepared by dissolving 1000 mg of metformin in 100 mL of methanol, yielding a solution of 1×10^7 ng/mL metformin.

4.4.6. Preparation of Calibration Curve

To obtain ten spiked levels (for the calibration curve) in plasma, the stock solution was diluted in methanol to obtain the following calibration curve concentrations: 2.5, 5, 14, 16, 19, 30, 33, 45, 60, and 112.5 μg (10,000 ng/mL).

4.5. Method Validation

4.5.1. Precision, Accuracy, and Absolute Recovery

The method's accuracy and precision were calculated by testing six samples, each with three independent replicates, on two days. The standard deviation (SD)-to-mean ratios were used to calculate relative standard deviation values (RSD) or CV%, expressed as percentages. The CV% limits for concentration and accuracy were determined to be less than 1.5. Furthermore, the agreed-upon accuracy criterion of 85–115% for all concentrations was achieved.

The absolute recovery of empagliflozin was determined using plasma samples prepared for drug or internal standard concentrations. The accuracy of the recovery should be consistent, exact, and replicable, with a maximum accuracy of 15% of the variance coefficient (percentage of CV).

4.5.2. Pharmacokinetic Analysis

Pharmacokinetic parameters were measured using the Winnonlin software V5.1, using a non-compartmental analysis (NCA) model. Estimations of the following parameters were made:

AUC_{last}: area under the curve to 96 h. AUC INF: area under the curve to infinity. C_{max}: maximum concentration of the drug in plasma. T_{max}: time to achieve C_{max}. $t_{1/2}$: elimination half-life. Kel: elimination rate constant.

4.5.3. Statistical Analysis

The statistical significance of the variable mean difference between the three groups, C_{max}, T_{max}, and AUC_{last}, was calculated using GraphPad Prism software 10 (GraphPad Software Inc.; San Diego, CA, USA) and an independent *t*-test sample—significant *p*-value < 0.05.

5. Conclusions

To conclude, our results showed the complex effect of grapefruit consumption in high amounts on the pharmacokinetic parameters of empagliflozin, including both the absorption rate and the extent of bioavailability, and the final profile in diabetic rats may be the results of compensatory mechanisms associated with the administration of a high amount of this drug in rats. This might raise an alarm in taking this drug concurrently with grapefruit juice in humans.

Future studies are required to validate our results further. These involve using both in vitro and in vivo models, in which such findings may have severe complications and side effects on patients and may lead to a loss of drug activity.

Author Contributions: Conceptualization, W.A.D. and M.H.; Data curation, W.A.D., M.H., Z.H. and H.K.N.; Formal analysis, W.A.D., S.M.A. and M.H.; Funding acquisition, W.A.D., M.H., N.M.A.-T., S.A., S.M.A. and R.A.; Investigation, W.A.D., S.M.A. and M.H.; Methodology, W.A.D., Z.Z., M.H., H.K.N., A.K., Z.H. and R.A.; Project administration, W.A.D. and M.H.; Resources, W.A.D. and M.H.; Software, W.A.D., Z.Z., M.H., N.M.A.-T., S.M.A. and S.A.; Supervision, W.A.D. and M.H.; Validation, W.A.D. and M.H.; Visualization, W.A.D. and M.H.; Writing—original draft, W.A.D., M.H. and A.K.; Writing—review and editing, W.A.D. and M.H. All authors have read and agreed to the published version of the manuscript.

Funding: This research received no external funding.

Institutional Review Board Statement: The animal study protocol was approved by the Institutional Review Board of Faculty of Pharmacy/Mutah University (protocol code SREC1132023, date 13 April 2023).

Informed Consent Statement: Not applicable.

Data Availability Statement: All the data are included in this article.

Acknowledgments: The authors sincerely thank all those who allowed us to complete this research successfully for their tremendous support.

Conflicts of Interest: The authors declare no conflicts of interest.

References

1. Sweidan, N.I.; Abu Khalaf, R.A.; Shatat, A.M.; Hammad, W.A. Therapeutic Potential of Silybum Marianum and Pergularia Tomentosa Extracts from Jordanian Origin in Diabetes Mellitus. *Curr. Bioact. Compd.* **2022**, *18*, 64–71. [CrossRef]

2. Harreiter, J.; Roden, M. Diabetes Mellitus: Definition, Classification, Diagnosis, Screening and Prevention (Update 2023). *Wien. Klin. Wochenschr.* **2023**, *135*, 7–17. [CrossRef]
3. Dostalek, M.; Akhlaghi, F.; Puzanovova, M. Effect of Diabetes Mellitus on Pharmacokinetic and Pharmacodynamic Properties of Drugs. *Clin. Pharmacokinet.* **2012**, *51*, 481–499. [CrossRef]
4. Chawla, G.; Chaudhary, K.K. A Complete Review of Empagliflozin: Most Specific and Potent SGLT2 Inhibitor Used for the Treatment of Type 2 Diabetes Mellitus. *Diabetes Metab. Syndr. Clin. Res. Rev.* **2019**, *13*, 2001–2008. [CrossRef]
5. Hailat, M.; Zakaraya, Z.; Al-Ani, I.; Al Meanazel, O.; Al-Shdefat, R.; Anwer, M.K.; Saadh, M.J.; Dayyih, W.A. Pharmacokinetics and Bioequivalence of Two Empagliflozin, with Evaluation in Healthy Jordanian Subjects under Fasting and Fed Conditions. *Pharmaceuticals* **2022**, *15*, 193. [CrossRef] [PubMed]
6. Hsia, D.S.; Grove, O.; Cefalu, W.T. An Update on SGLT2 Inhibitors for the Treatment of Diabetes Mellitus. *Curr. Opin. Endocrinol. Diabetes Obes.* **2017**, *24*, 73. [CrossRef] [PubMed]
7. Steiner, S. Empagliflozin, Cardiovascular Outcomes, and Mortality in Type 2 Diabetes. *Z. Gefassmedizin* **2016**, *13*, 17–18.
8. Briguglio, M.; Hrelia, S.; Malaguti, M.; Serpe, L.; Canaparo, R.; Dell'osso, B.; Galentino, R.; De Michele, S.; Dina, C.Z.; Porta, M.; et al. Food Bioactive Compounds and Their Interference in Drug Pharmacokinetic/Pharmacodynamic Profiles. *Pharmaceutics* **2018**, *10*, 277. [CrossRef] [PubMed]
9. Awad, R.; Mallah, E.; Alkhawaja, B.; Abudayyih, W.; El-Hajji, F.; Matalka, K.Z.; Arafat, T. Pomegranate and Licorice Juices Modulate Metformin Pharmacokinetics in Rats. *Neuroendocrinol. Lett.* **2016**, *37*, 202–206.
10. Bailey, D.G. Fruit Juice Inhibition of Uptake Transport: A New Type of Food-Drug Interaction. *Br. J. Clin. Pharmacol.* **2010**, *70*, 645–655. [CrossRef]
11. Fares, M.; Al-Ani, I.; Hailat, M.; Khaleel, A.; Collier, P.; Al-Ani, A.; Abu Dayyih, W. Evaluation the Impact of Grapefruit Juice on Sitagliptin Pharmacokinetics in Healthy Rats. *Int. J. Pharm. Res.* **2021**, *13*. [CrossRef]
12. Santes-Palacios, R.; Romo-Mancillas, A.; Camacho-Carranza, R.; Espinosa-Aguirre, J.J. Inhibition of Human and Rat CYP1A1 Enzyme by Grapefruit Juice Compounds Rafael. *Toxicol. Lett.* **2016**, *258*, 267–275. [CrossRef]
13. Trang, N.N.; Chung, C.-C.; Lee, T.-W.; Cheng, W.-L.; Kao, Y.-H.; Huang, S.-Y.; Lee, T.-I.; Chen, Y.-J. Empagliflozin and Liraglutide Differentially Modulate Cardiac Metabolism in Diabetic Cardiomyopathy in Rats. *Int. J. Mol. Sci.* **2021**, *22*, 1177. [CrossRef]
14. Schwaiger, E.; Burghart, L.; Signorini, L.; Ristl, R.; Kopecky, C.; Tura, A.; Pacini, G.; Wrba, T.; Antlanger, M.; Schmaldienst, S.; et al. Empagliflozin in Posttransplantation Diabetes Mellitus: A Prospective, Interventional Pilot Study on Glucose Metabolism, Fluid Volume, and Patient Safety. *Am. J. Transplant.* **2019**, *19*, 907–919. [CrossRef] [PubMed]
15. Jojima, T.; Sakurai, S.; Wakamatsu, S.; Iijima, T.; Saito, M.; Tomaru, T.; Kogai, T.; Usui, I.; Aso, Y. Empagliflozin Increases Plasma Levels of Campesterol, a Marker of Cholesterol Absorption, in Patients with Type 2 Diabetes: Association with a Slight Increase in High-Density Lipoprotein Cholesterol. *Int. J. Cardiol.* **2021**, *331*, 243–248. [CrossRef]
16. Patoulias, D.; Papadopoulos, C.; Doumas, M. Effect of Empagliflozin on Cholesterol Synthesis and Absorption Markers in Patients with Type 2 Diabetes: Any Role of DPP-4 Inhibitors? *Int. J. Cardiol.* **2021**, *330*, 228. [CrossRef] [PubMed]
17. Di Marco, M.P.; Edwards, D.J.; Wainer, I.W.; Ducharme, M.P. The Effect of Grapefruit Juice and Seville Orange Juice on the Pharmacokinetics of Dextromethorphan: The Role of Gut CYP3A and P-Glycoprotein. *Life Sci.* **2002**, *71*, 1149–1160. [CrossRef] [PubMed]
18. Ahmed, I.S.; Hassan, M.A.; Kondo, T. Effect of Lyophilized Grapefruit Juice on P-Glycoprotein-Mediated Drug Transport in-Vitro and in-Vivo. *Drug Dev. Ind. Pharm.* **2015**, *41*, 375–381. [CrossRef] [PubMed]
19. Shen, X.; Chen, F.; Wang, F.; Huang, P.; Luo, W. The Effect of Grapefruit Juice on the Pharmacokinetics of Tadalafil in Rats. *Biomed. Res. Int.* **2020**, *2020*, 1631735. [CrossRef] [PubMed]
20. Sattar, N.; Fitchett, D.; Hantel, S.; George, J.T.; Zinman, B. Empagliflozin Is Associated with Improvements in Liver Enzymes Potentially Consistent with Reductions in Liver Fat: Results from Randomised Trials Including the EMPA-REG OUTCOME® Trial. *Diabetologia* **2018**, *61*, 2155–2163. [CrossRef] [PubMed]
21. Lv, Q.; Le, L.; Xiang, J.; Jiang, B.; Chen, S.; Xiao, P. Liver Transcriptomic Reveals Novel Pathways of Empagliflozin Associated with Type 2 Diabetic Rats. *Front. Endocrinol.* **2020**, *11*, 111. [CrossRef] [PubMed]
22. Pingili, R.; Vemulapalli, S.; Mullapudi, S.S.; Nuthakki, S.; Pendyala, S.; Kilaru, N. Pharmacokinetic Interaction Study between Flavanones (Hesperetin, Naringenin) and Rasagiline Mesylate in Wistar Rats. *Drug Dev. Ind. Pharm.* **2016**, *42*, 1110–1117. [CrossRef] [PubMed]
23. Alkather, Z.; Hailat, M.; Al-Shdefat, R.; Abu Dayyih, W. Development and Validation of HPLC Method for Five Gliptins in Pharmaceutical Dosage Forms in Finished Marketed Products. *Curr. Pharm. Anal.* **2020**, *17*, 1263–1271. [CrossRef]

molecules

MDPI

Article

Analysis of Softwood Lignans by Comprehensive Two-Dimensional Liquid Chromatography

Danil I. Falev *, Ilya S. Voronov , Alexandra A. Onuchina , Anna V. Faleva , Nikolay V. Ul'yanovskii * and Dmitry S. Kosyakov

Laboratory of Natural Compounds Chemistry and Bioanalytics, Core Facility Center "Arktika", M.V. Lomonosov Northern (Arctic) Federal University, Northern Dvina Emb. 17, 163002 Arkhangelsk, Russia; i.voronov@narfu.ru (I.S.V.); a.onuchina@narfu.ru (A.A.O.); a.bezumova@narfu.ru (A.V.F.); d.kosyakov@narfu.ru (D.S.K.)

* Correspondence: d.falev@narfu.ru (D.I.F.); n.ulyanovsky@narfu.ru (N.V.U.)

Abstract: Lignans constitute a large group of phenolic plant secondary metabolites possessing high bioactivity. Their accurate determination in plant extracts with a complex chemical composition is challenging and requires advanced separation techniques. In the present study, a new approach to the determination of lignans in coniferous knotwood extracts as the promising industrial-scale source of such compounds based on comprehensive two-dimensional liquid chromatography separation and UV spectrophotometric detection is proposed. First and second-dimension column screening showed that the best results can be obtained using a combination of non-polar and polar hydroxy group embedded octadecyl stationary phases with moderate (~40%) "orthogonality". The optimization of LC × LC separation conditions allowed for the development of a new method for the quantification of the five lignans (secoisolariciresinol, matairesinol, pinoresinol, 7-hydroxymatairesinol, and nortrachelogenin) in knotwood extracts with limits of quantification in the range of 0.27–0.95 mg L^{-1} and a linear concentration range covering at least two orders of magnitude. Testing the developed method on coniferous (larch, fir, spruce, and pine) knotwood extracts demonstrated the high selectivity of the analysis and the advantages of LC × LC in the separation and accurate quantification of the compounds co-eluting in one-dimensional HPLC.

Keywords: comprehensive two-dimensional liquid chromatography; LC × LC; lignans; coniferous knotwood

1. Introduction

Among the polyphenolic secondary metabolites of plants, a special place is occupied by lignans, which contain two phenylpropane units connected by the β-β alkyl-alkyl or other types of bonds (alkyl-alkyl and alkyl-aryl) in their structure. In the latter case, it is customary to use the term neolignans [1–3]. Lignans and neolignans are extremely widespread in nature and number more than a thousand currently known representatives, and this list is constantly expanding. Due to the unique biological activity of lignans possessing antioxidant, antitumor, hepatoprotective, and cardioprotective properties, the search for their natural sources and the development of analytical methods for the determination of these compounds in plant raw materials are becoming increasingly important.

Currently, compression wood (knots, roots) of coniferous trees, the active biosynthesis of lignans in which is a response to mechanical stress, is considered one of the most promising industrial sources of such compounds [4,5]. It is known that coniferous knotwood may contain up to 20% lignans, and their major representatives are 7-hydroxymatairesinol or HMR (spruce), nortrachelogenin (pine), and secoisolariciresinol (larch, fir). Matairesinol and pinoresinol are also present in large amounts [6]. In addition to the above-mentioned bioactive properties, these compounds are capable of being metabolized under the action of the gut microbiome with the formation of so-called enterolignans or mammalian lignans (enterodiol and enterolactone), which play an important biological role. In addition to these components, coniferous wood contains a number of other lignans, extractive substances of different classes (flavonoids, steroids, resin acids, etc.), and their glycosylated derivatives. This complicates the detailed analysis of the compression wood extracts and necessitates the use of advanced analytical techniques.

Currently, the qualitative and quantitative analysis of lignans is carried out mainly by thin-layer (TLC), gas (GC), and high-performance liquid chromatography (HPLC), their combinations with various types of mass spectrometry (MS), and matrix-assisted laser desorption/ionization mass spectrometry (MALDI). TLC has mainly been used as an inexpensive adjuvant method in the study of various plant materials, including coniferous wood [7,8]. The high-performance version of this separation technique (HPTLC) was successfully used for the isolation, identification, and quantitative assessment of representatives of lignans, such as schisandrol A and schisandrol B, in various preparations of Chinese lemongrass (*Schisandra chinensis*) [9], phyllanthin, hypophyllanthin, niranthin, and nirtetralin in Phyllanthus species [10] as well as sesamin and sesamolin in sesame oil and its polyherbal formulations [11]. GC provides higher efficacy in lignan separation and is extensively used for identification and quantification purposes [12,13] in combination with MS detection. However, due to the low volatility and thermal lability of lignans, GC-MS requires preliminary derivatization of the analytes (typically, silylation) [14], which is a tedious time and labor consuming procedure that negatively affects the accuracy and reproducibility of the analysis. MALDI MS is distinguished by the simplicity of the sample preparation and tolerance to impurities; however, it cannot be easily combined with separation techniques [15].

In this regard, HPLC allowing the direct (without derivatization step) analysis of plant extracts can be considered a method of choice for the detection, identification, and quantification of lignans. The separation of the analytes is typically carried out on reversed stationary phases (e.g., octadecyl silica, C18) in gradient elution mode, providing sufficient resolution towards the major components of plant extracts. Thus, the authors of [16,17] reported the separation of isolariciresinol, secoisolariciresinol, anhydrosecoisolariciresinol, matairesinol, lariciresinol, hinokinin, arctigenin, and pinoresinol within 50 min and the possibility to resolve enantiomeric compounds using chiral stationary phases. In other noteworthy studies [18,19], the separation of ten Chinese lemongrass lignans was achieved within 40 min.

Various types of detection are used in combination with HPLC in lignan analysis, while UV spectrophotometry [7,16–19] and MS [6,20,21] are the most common techniques. The latter provides the highest selectivity and sensitivity and ensures the reliable identification of unknown compounds. At the same time, the quantification of analytes by HPLC-MS (especially with electrospray ionization that is extremely susceptible to matrix effects) requires corresponding analytical standards, which are hardly commercially available in the case of lignans. Apparently, HPLC-UV is less expensive and provides more possibilities for the standard-free semi-quantification of structurally close analytes containing the same chromophores due to their similar response factors and lower matrix effects in spectrophotometry when compared to MS. However, overcoming the low selectivity of UV detection requires achieving the highest chromatographic resolution between all the analytes and matrix components, which is difficult to achieve for such complex samples as plant extracts [7,17], even in ultra-performance (UPLC) analysis.

In our opinion, this issue can be resolved by implementing comprehensive two-dimensional liquid chromatography (LC $\times$ LC), which involves the consecutive separation of all the sample components on "orthogonal" (in chemical nature) first- (^{1}D) and second-dimension (^{2}D) columns [22]. Due to its high separation efficiency and peak capacity [23], LC $\times$ LC is increasingly used in the analysis of natural compounds, particularly plant secondary metabolites [22]. Examples include the determination of phenolic acids in wine [24] and flavonoids in plant extracts [25]. Specific lignans of Chinese lemongrass, schisandrins, were also successfully separated by LC $\times$ LC [26]. Despite this, there is still no information in the literature about the possibility of using LC × LC for the highly efficient analysis of softwood lignans.

The present study aimed to fill this gap and focused on developing an approach to the quantitative determination of lignans in coniferous knotwood extracts as the most important industrial-scale source of these compounds based on a combination of comprehensive two-dimensional liquid chromatography with diode array spectrophotometric detection.

2. Results and Discussion

2.1. Column Screening and Selection of LC × LC Conditions

A key factor in the LC × LC method development is the proper selection of a combination of stationary phases that ensures the most complete separation of the analytes. Considering the existing limitations on the composition (and thus elution power) of the sample solvent introduced into the ^{2}D column, combinations of stationary phases with completely different retention mechanisms (for example, reversed and normal phase or hydrophilic interaction retention) were not used in our study. Instead, five reversed stationary phases, which have previously been proven to be effective in lignan separations [16–19], were chosen for further testing. They differ in the presence and nature of the embedded polar functional groups that affect the separation selectivity: (i) Shim-pack XR-ODS II—octadecyl-bonded silica, endcapped; (ii) Nucleodur C18 Isis—cross-linked octadecyl-bonded silica, endcapped; (iii) Nucleodur C18 Pyramid—octadecyl-bonded silica with hydrophilic (-CH_2OH) endcapping; (iv) Nucleodur PolarTec—octadecyl-bonded silica with embedded polar (amide) groups, endcapped; and (v) Nucleodur PFP—pentafluorophenyl propyl-bonded silica, partially endcapped.

Considering the need to separate not only the most important lignans but also the matrix components affecting the target analytes' UV detection, chromatographic runs were performed using the larch knotwood extract as a test sample. The six combinations of stationary phases (Table 1), including three pairs of ^{1}D non-polar (Shim-pack XR-ODS II, Nucleodur Isis) and ^{2}D polar functionalized (Nucleodur C18 Pyramid, Nucleodur PFP), and three pairs of ^{1}D and ^{2}D polar functionalized sorbents, were tested (Supplementary Figure S1).

Table 1. Orthogonality (A_O) and regression coefficient (R^2) parameters for different combinations of stationary phases in the LC × LC separation of lignan-rich larch knotwood extract.

Stationary Phase		A_O (%)	R^2
^{1}D	^{2}D		
Shim-pack XR-ODS II	Nucleodur C18 Pyramid	39	0.84
Nucleodur C18 Isis	Nucleodur C18 Pyramid	34	0.92
Nucleodur PFP	Nucleodur C18 Pyramid	40	0.85
Nucleodur PolarTec	Nucleodur PFP	38	0.88
Shim-pack XR-ODS II	Nucleodur PFP	38	0.87
Nucleodur PolarTec	Nucleodur C18 Pyramid	36	0.88

The stationary phase orthogonality parameters (A_O) and regression coefficient (R^2) calculated from the measured retention times using "Asterisk" equations according to Camenzuli and Schoenmakers [27] are presented in Table 1. Since for the most efficient separation, the A_0 and R^2 values should be as high and as low as possible, respectively [28], the best results were observed for the following two combinations of stationary phases: ^{1}D Shim-pack XR-ODS II—^{2}D Nucleodur C18 Pyramid (Supplementary Figure S1a) and ^{1}D Nucleodur PFP—^{2}D Nucleodur C18 Pyramid (Supplementary Figure S1b). The attained orthogonality values for these pairs (39 and 40%, respectively) can be considered quite acceptable given that A_O > 43% already refers to high orthogonality [29]. Although both the A_O and R^2 values for all the tested stationary phase combinations fell into rather narrow ranges (34–40% and 0.84–0.92, respectively) and, thus, good separation was achieved in all cases, the advantages of the two selected column pairs can be noticed, even visually, in Supplementary Figure S1. Considering the higher availability of the octadecyl stationary phases, the combination of the Shim-pack XR-ODS II ([1]D) and Nucleodur C18 Pyramid ([2]D) chromatographic columns was chosen for further studies.

Variation of the gradient elution profiles in both the [1]D and [2]D dimensions allowed the establishment of the elution programs with water (A) and acetonitrile (B), both acidified with 0.1% of formic acid, as mobile phase components: [1]D—start from 15% B with linear ramp to 65% B during 60 min; [2]D—start from 20% B with linear ramp to 90% B during 0.75 min, 20% B from 0.75 to 1.00 min for column equilibration. They ensured the achievement of the maximum distribution of the detected compounds in the 2D chromatogram (Supplementary Figure S1a), their fast elution (within 1 min) from the [2]D column, and the full separation of lignans with similar retention on the reversed-phase columns—HMR and secoisolariciresinol (Figure 1). Since the maximum absorption of all the tested lignans was observed at 280 nm (Figure S2), the construction of two-dimensional chromatographic plots and the quantification were carried out at that wavelength.

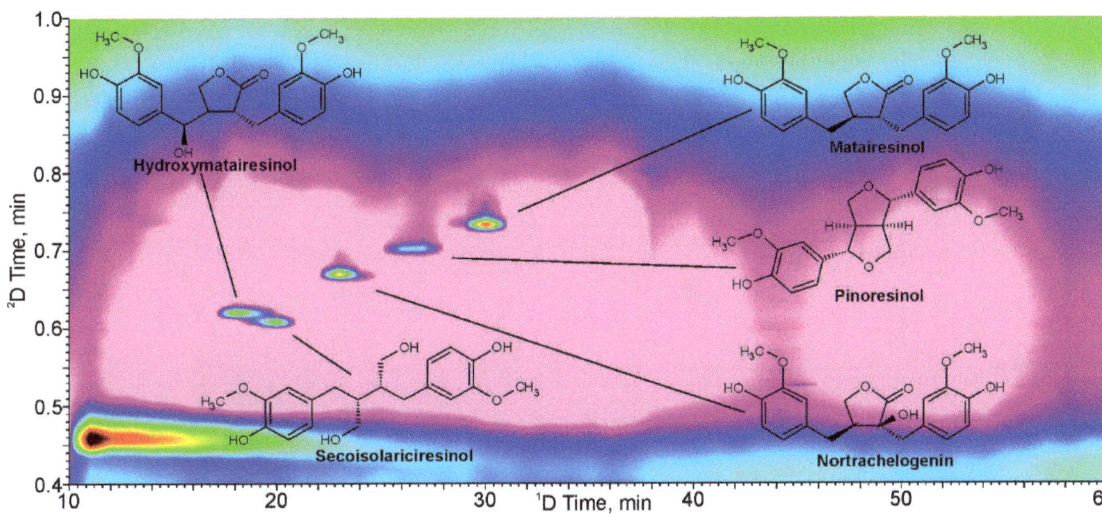

Figure 1. LC × LC-UV chromatogram (280 nm) of analytes model mixture (10 mg L^{-1}) and chemical structures of the studied lignans.

2.2. *Quantitative Analysis and Method Validation*

For the quantitative method development, five major representatives of coniferous wood lignans (secoisolariciresinol, matairesinol, pinoresinol, HMR, and nortrachelogenin) found in the majority of knotwood samples [6] were chosen as the target analytes. The analyses of the model solutions of various concentrations (up to 20 mg L^{-1}) with the further construction of calibration plots (chromatographic peak area S vs. concentration C) showed good linearity ($R^2 > 0.999$) in the range of at least two orders of magnitude (Table 2). Rather close response factors (calibration line slope *a*) differing by a maximum of two times were observed for all the analytes. The instrumental limits of detection (LODs) and quantification (LOQs) determined as the analyte concentrations providing signal-to-noise ratios of 3:1 and 10:1, respectively, were in the ranges of 0.08–0.29 (LOD) and 0.27–0.95 (LOQ) mg L^{-1}. These values were additionally confirmed in the analysis of the sample with analyte concentrations close to the LOQ (Supplementary Figure S3). The attained sensitivity level is typical for LC-UV, and the obtained LOQs turned out to be noticeably lower than those reported in [11] for schisandrins (0.67 to 4.83 mg L^{-1}).

Table 2. Calibration dependences (S = aC + b) for the area of chromatographic peak versus analyte concentration, LODs, LOQs, and retention times of analytes.

Analyte	Retention Time, min		Linear Range, mg L^{-1}	a	b	R^2	LOD, mg L^{-1}	LOQ, mg L^{-1}
	^1D	^2D						
HMR	18.6	0.62	0.45–20	15389	−4511	>0.999	0.13	0.44
Secoisolariciresinol	20.6	0.61	0.35–20	20407	−5854	>0.999	0.16	0.54
Nortrachelogenin	23.7	0.67	0.36–20	16557	−1944	>0.999	0.11	0.37
Pinoresinol	27.7	0.70	0.29–20	37764	−10824	>0.999	0.29	0.95
Matairesinol	30.7	0.73	0.31–20	21436	−2998	>0.999	0.08	0.27

As can be seen from Table 3 containing the results of intra- and inter-day assays, the achieved accuracy of the developed method was in the range of 91–109%, while the standard deviation (precision) did not exceed 13% at the LOQ level. The matrix effects were estimated by a spike recovery test using birch xylem extract as a matrix, which does not contain coniferous lignans. The obtained recovery values ranged from 82 to 98% for two levels of analyte concentrations (close to LOQ and 10 LOQ), indicating no substantial matrix interferences for all the analytes (Table 4). The effective elimination of the matrix effect was achieved by two-dimensional chromatographic separation of lignans from the matrix.

Table 3. Method accuracy and precision estimated in intra-day and inter-day assays of the model solution of lignans.

Analyte	Concentration, mg L^{-1}	Intra-Day Assay (n = 6)			Inter-Day Assay (n = 6)		
		Found, mg L^{-1}	Accuracy, %	Precision, %	Found, mg L^{-1}	Accuracy, %	Precision, %
HMR	0.50	0.46 ± 0.03	91	4.66	0.45 ± 0.02	90	3.14
Secoisolariciresinol	0.50	0.51 ± 0.08	102	11.1	0.46 ± 0.04	92	6.11
Nortrachelogenin	0.50	0.49 ± 0.05	97	7.29	0.45 ± 0.07	89	11.1
Pinoresinol	1.00	1.00 ± 0.02	99	1.43	0.51 ± 0.09	101	12.6
Matairesinol	0.50	0.54 ± 0.07	107	9.25	0.55 ± 0.07	109	9.08

Table 4. Matrix effect estimated by the spike recovery test.

Analyte	Spiked, mg L^{-1}	Found, mg L^{-1}	Recovery, %
HMR	1.0	0.82 ± 0.06	82
	10	8.4 ± 0.8	84
Secoisolariciresinol	1.0	0.82 ± 0.15	82
	10	9.0 ± 0.4	90
Nortrachelogenin	1.0	0.85 ± 0.08	85
	10	9.3 ± 0.7	93
Pinoresinol	1.0	8.8 ± 0.7	88
	10	9.8 ± 0.2	98
Matairesinol	1.0	8.9 ± 1.0	89
	10	9.0 ± 0.4	90

2.3. *Analysis of Coniferous Knotwood Extracts*

To test the developed approach, larch, fir, spruce, and pine knotwood acetone extracts were selected as the real samples. These objects are characterized by a complex chemical composition and contain a number of lignans. The obtained 2D chromatograms (Figure 2) demonstrate the presence of all the target analytes in a wide concentration range as well as other numerous components absorbing UV radiation (No. 1–19). The comparison of the one- and two-dimensional HPLC-UV chromatograms obtained on the same ^1D column (Shim-pack XR-ODS II) clearly demonstrates the undoubted advantages of LC × LC.

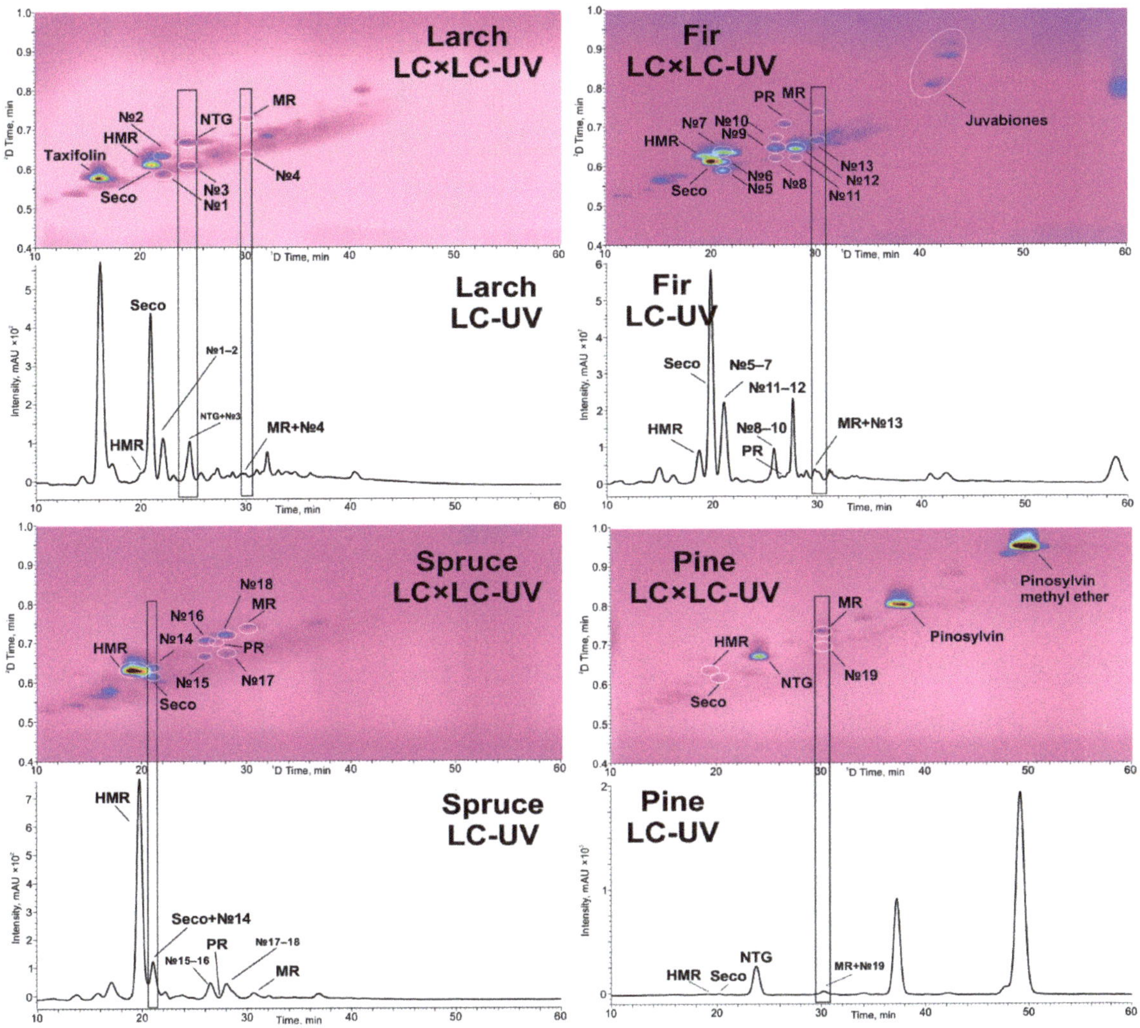

Figure 2. LC × LC-UV and LC-UV chromatograms (280 nm) of softwood knots extracts (MR—matairesinol, Seco—secoisolariciresinol, PR—pinoresinol, NTG—nortrachelogenin).

The latter made it possible to separate the target analytes from other co-eluting in ^{1}D compounds, for example, nortrachelogenin and unidentified compound No. 2, matairesinol and No. 4, matairesinol and No. 13, secoisolariciresinol and No. 14, and matairesinol and No 19. Moreover, the unidentified compounds, many of which belong to the lignan family, were also well separated (for example, No. 1–2; 5–7; 8–10; 11–12; 15–16; and 17–18). Despite the longer separation time (60 min), the LC × LC-UV approach has obvious advantages, even over LC-MS, both in terms of the cost of analysis and the elimination of matrix effects. The latter factor, together with similar absorption coefficients of various lignans in the UV region, can be the basis for the development of methods for the semi-quantitative standard-free determination of such compounds in plant extracts.

The results of the quantification of the target analytes (Table 5) are consistent with the literature data [6,30] and demonstrate the predominance of HMR (100 mg g^{-1}), secoisolariciresinol (20 mg g^{-1}), and nortrachelogenin (8 mg g^{-1}) in spruce, fir and larch, and pine knotwood, respectively.

Table 5. The content of lignans (mg g^{-1}, recalculated for the oven-dried plant material) in softwood knots (n = 2, p = 0.95).

Analyte	Larch	Fir	Spruce	Pine
HMR	1.2 ± 0.2	0.17 ± 0.04	100 ± 10	0.23 ± 0.03
Secoisolariciresinol	17 ± 6	20 ± 2	6.1 ± 0.7	0.24 ± 0.08
Nortrachelogenin	3.3 ± 0.4	-	-	8.0 ± 0.9
Pinoresinol	-	0.78 ± 0.14	0.31 ± 0.05	-
Matairesinol	0.55 ± 0.07	0.75 ± 0.17	1.0 ± 0.2	0.65 ± 0.13

In addition to lignans, other phenolic compounds were also found in the studied samples and were tentatively identified on the basis of data in the literature [5,6], UV absorption spectra (Supplementary Figures S4–S6) [31–33], and retention times. These include taxifolin (^{1}D 16.05 min, ^{2}D 0.60 min) in larch, pinosylvin (^{1}D 37.60 min, ^{2}D 0.80 min), its methyl ether (^{1}D 49.80 min, ^{2}D 0.95 min) in pine, and three juvabiones (^{1}D 41.05 min, ^{2}D 0.81 min; ^{1}D 43.20 min, ^{2}D 0.88 min; ^{1}D 43.20 min, ^{2}D 0.91 min) in fir knotwood extracts.

3. Materials and Methods

3.1. Chemicals and Reagents

Commercially available standards of pinoresinol (≥95%), secoisolariciresinol (≥95%), and matairesinol (≥85%) were purchased from Sigma-Aldrich (St. Louis, MO, USA). 7-hydroxymatairesinol (≥98%) and nortrachelogenin (≥98%), which are poorly available as high-purity commercial preparations, were obtained in our laboratory from knotwood extracts by preparative liquid chromatography (Section 3.3).

Acetonitrile (HPLC gradient grade, Khimmed, Saint Petersburg, Russia), formic acid (≥96%, Sigma-Aldrich, St. Louis, MO, USA), and "type I" Milli-Q high-purity water were used for the preparation of the mobile phase. Methanol (high-purity grade, Khimmed, Saint Petersburg, Russia) was used for the preparation of the sample solutions. Hexane (chem. pure grade, Khimmed, Saint Petersburg, Russia) and acetone (pure, Vershina, Vsevolozhsk, Russia) were used for Soxhlet extraction.

The stock solutions of lignans in methanol (500 mg L^{-1}) were prepared from accurately weighed samples and stored at 4 °C for no longer than one week. The model and calibration solutions of the analytes were prepared immediately before the analyses by mixing and successive dilutions of the stock solutions with methanol.

3.2. Plant Materials and Extraction

Four coniferous tree species were chosen as a source of plant material in our study—Norway spruce (*Picea abies*), Siberian fir (*Abies sibirica*), Scotch pine (*Pinus sylvestris*), and Larch (*Larix sibirica*). The tree trunks were harvested in boreal forests of the European North of Russia in the following locations: 64°76′ N 40°80′ E (pine), 64°28′ N 40°76′ E (spruce) 61°77′ N 42°47′ E (fir), and 61°16′ N 42°54′ E (larch). The parts with large knots were cut from at least five trunks of each species and immediately delivered to the laboratory. The knotwood samples were taken from the inner (inside the trunk) part of the knots by drilling and were carefully averaged. The obtained knotwood shavings were vacuum dried at 40 °C overnight and then milled in a ZM 200 centrifugal mill (Retsch, Haan, Germany) to a particle size of <1 mm.

Soxhlet extraction of the prepared sawdust was carried out according to a previously developed procedure [6]. The dried material (~10 g) was extracted using hexane for 8 h to remove lipids and resin. After air-drying, the plant material was subjected to exhaustive extraction using acetone for 8 h. The obtained acetone extracts were evaporated on a rotary evaporator RV 10 (IKA, Königswinter, Germany) to dryness. The attained yields of the extractive substances were larch—11.2%, fir—16.8%, spruce—16.3%, and pine—3.85%.

The birch (*Betula pendula*) xylem extract was used for the estimation of the matrix effects (hardwood is supposed to be lignan-free).

3.3. Isolation and Characterization of Hydroxymatairesinol and Nortrachelogenin

Since 7-hydroxymatairesinol (HMR) and nortrachelogenin sharply predominate in spruce and pine knotwood extracts and can be effectively separated from matrix compounds [34], their preparations of sufficient purity were obtained by preparative HPLC. Chromatographic separations were carried out at a temperature of 40 °C using a semipreparative-scale Nucleodur C18 Gravity column (Macherey-Nagel, Duren, Germany), 250 × 21 mm, 5 μm particle size, on an LC-20 preparative HPLC system (Shimadzu, Kyoto, Japan) consisting of two LC-20AP pumps with high-pressure gradient formation, a vacuum degasser, a CTO-20A column thermostat, an SPD-M20A high-flow diode array UV-VIS detector, and an FRC-10A fraction collector. The system was controlled using LabSolutions software v. 5.54 (Shimadzu, Kyoto, Japan). The mobile phase was a mixture of components A (0.1% aqueous solution of formic acid) and B (acetonitrile with 0.1% of formic acid) with a total flow rate of 21.0 mL min^{-1}. The following gradient elution program was applied: 0–20 min—25% B; 20–25 min—linear ramp to 100% B, held for 10 min. The total separation time was 35 min.

The 50 mg samples of the dried knotwood extracts were redissolved in 2 mL of 50% aqueous methanol, centrifuged, and manually injected (2 mL) into the HPLC system. The target fractions of HMR and nortrachelogenin were collected in the periods of 9.50–10.50 and 16.20–17.50 min, respectively (Supplementary Figure S7) and then evaporated on a rotary evaporator to dryness and redissolved in 1 mL of methanol. Their purity was estimated in an additional chromatographic analysis using the same analytical HPLC system as for the lignan analysis (Section 3.4). The content of the target compound (%) was calculated as a ratio of the corresponding chromatographic peak area and the total area of all the peaks on the chromatogram.

The structures of the isolated individual compounds were confirmed by their ^{1}H and ^{13}C NMR spectra registered at 25 °C in CD_3OD on an AVANCE III NMR spectrometer (Bruker, Ettlingen, Germany), with an operational frequency of 600 MHz (^{1}H). The following parameters were applied: (i) ^{1}H NMR: zg30 sequence, acquisition time—1.4 s, relaxation delay—1 s, number of data points—64,000, number of scans—8, spectrum window width ~15 ppm; (ii) ^{13}C NMR: zgig30 sequence, pulse width—12 ms; relaxation delay—2 s; number of data points—64,000, number of scans—1024, spectrum window width ~240 ppm. The solvent signal was used as an internal standard (dC/dH 49.15/3.31 ppm).

The characteristics of the obtained preparations were as follows. 7-Hydroxymatairesinol: light beige powder; purity: 98%; UV (MeOH): λ max 228, 280 nm; ^{1}H NMR (600 MHz, CD_3OD) and ^{13}C NMR (150 MHz, CD_3OD). The ^{1}H and ^{13}C NMR data are presented in Supplementary Tables S1 and S2, Figures S8 and S9.

Nortrachelogenin: light beige powder; purity: 98%; UV (MeOH): λ max 228, 280 nm; ^{1}H NMR (600 MHz, CD_3OD) and ^{13}C NMR (150 MHz, CD_3OD). The ^{1}H and ^{13}C NMR data are presented in Supplementary Tables S1 and S2, Figures S10 and S11.

3.4. Comprehensive Two-Dimensional Liquid Chromatography

A comprehensive two-dimensional liquid chromatography system, Nexera-e (Shimadzu, Kyoto, Japan), was used for the one- and two-dimensional analytical separations and consisted of four LC-30AD pumps with gradient formation on the high-pressure side, two five-channel degasser units DGU-A5, an LC-30AC autosampler, a CTO-30A column thermostat, two high-speed/high-pressure six-port switching valves, equipped with two sampling loops (volume 50 μL) and an SPD-M20A diode array detector. A schematic diagram of the LC × LC system is presented in Figure 3. The controlling of the HPLC system as well as the data collection and processing were carried out using the LabSolutions 5.65 software package (Shimadzu, Kyoto, Japan).

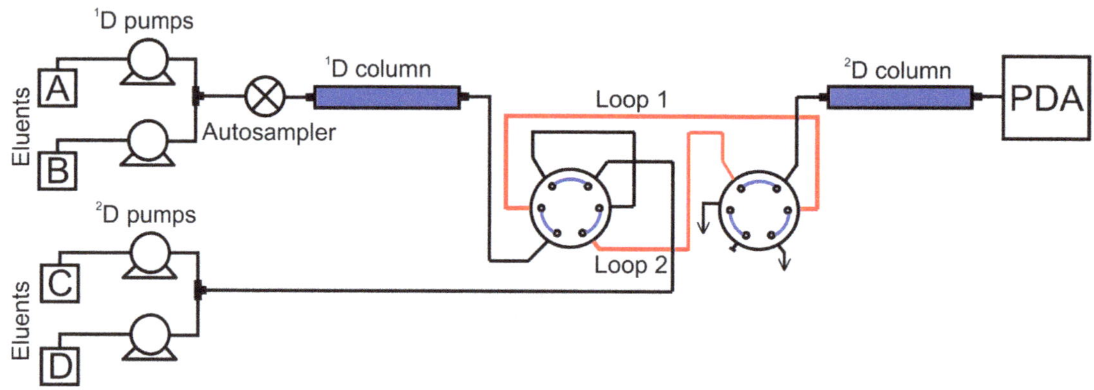

Figure 3. Schematic diagram of the LC × LC-UV system.

Knotwood extracts (1.00 mg) were dissolved in 1 mL of methanol, centrifuged, and injected into the LC × LC system. Separation was performed on the corresponding ^{1}D and ^{2}D columns with flow rates of 0.05 and 2.0 mL min^{-1}, respectively. The effluent from the first column was transferred to the loop of the flow-switching valve. Every 60 s (modulation time), the collected 50 μL portion of the ^{1}D column effluent was injected into the ^{2}D column by switching the valve. ChromSquare 2.2 (Chromaleont, Messina, Italy) was used for the system control and the construction of the LC × LC chromatograms. The chromatographic separations were carried out at 40 °C. The detection was performed at 280 nm. The injection volume was 4 μL.

The following chromatographic columns were used in the ^{1}D separations: Shim-pack XR-ODS II, 3 × 50 mm, 2.2 μm particle size (Shimadzu, Kyoto, Japan); Nucleodur C18 Isis, 2.0 × 150 mm, 1.8 μm particle size (Macherey-Nagel, Duren, Germany); Nucleodur PFP, 2.0 × 150 mm, 1.8 μm particle size (Macherey-Nagel, Duren, Germany); Nucleodur PolarTec, 2.0 × 150 mm, 1.8 μm particle size (Macherey-Nagel, Duren, Germany). In the ^{2}D separations, the following columns were used: Nucleodur PFP, 4.6 × 30 mm, 1.8 μm particle size (Macherey-Nagel, Duren, Germany) and Nucleodur C18 Pyramid, 4.6 × 50 mm, 1.8 μm particle size (Macherey-Nagel, Duren, Germany).

4. Conclusions

Comprehensive two-dimensional liquid chromatography on reversed stationary phases with different chemistries ensures the efficient separation of coniferous wood lignans in plant extracts in the gradient elution mode. The highest "orthogonality" and best separation are achieved on a combination of non-polar and polar hydroxy group embedded octadecyl stationary phases in the first and second dimensions, respectively. On this basis, a novel method for the quantification of the five lignans (secoisolariciresinol, matairesinol, pinoresinol, HMR, and nortrachelogenin) in knotwood extracts by LC × LC with UV diode array detection was developed and validated as an alternative to LC-MS assays. The attained LOQs of the target analytes ranged from 0.27 to 0.95 mg L^{-1}, and the calibration plots were linear in the concentration range covering at least two orders of magnitude. Testing the developed method on coniferous (larch, fir, spruce, and pine) knotwood extracts demonstrated the high selectivity of the analysis, which allowed the separation and quantification of the target analytes and matrix components co-eluting in one-dimensional HPLC. The proposed approach can be used to develop and improve methods for the determination of lignans in various objects. Further studies should be focused on the hyphenation of LC × LC separation with mass spectrometric detection for non-target screening, identification, and the highly sensitive determination of lignans and their derivatives in plant extracts.

Supplementary Materials: The following supporting information can be downloaded at https://www.mdpi.com/article/10.3390/molecules28248114/s1, Figure S1: LC × LC-UV chromatograms (280 nm) of larch knotwood extract recorded on different 1D and 2D column combinations; Figure S2: Scaled UV spectra of lignans; Figure S3: LC × LC-UV chromatogram (280 nm) of analyte model mixture (1.0 mg L^{-1}).; Figure S4: UV spectrum of the component of the larch knotwood extract with ^{1}D Time 16.05 min and ^{2}D Time 0.58 min; Figure S5: UV spectra of the components of pine knotwood extract; Figure S6: UV spectra of the components of fir knotwood extract; Figure S7: Preparative LC-UV chromatograms (280 nm) of extract; Figure S8: ^{1}H NMR spectrum of hydroxymatairesinol (Methanol-d_4); Figure S9: ^{13}C NMR spectrum of hydroxymatairesinol (Methanol-d_4); Figure S10: ^{1}H NMR spectrum of nortrachelogenin (Methanol-d_4); Figure S11: ^{13}C NMR spectrum of nortrachelogenin (Methanol-d_4). Table S1: ^{1}H NMR (600 MHz) Spectroscopic Data of hydroxymatairesinol and nortrachelogenin; Table S2: ^{13}C NMR (150 MHz) Spectroscopic Data of hydroxymatairesinol and nortrachelogenin.

Author Contributions: Conceptualization: D.I.F. and I.S.V.; Methodology: D.I.F., I.S.V., A.A.O., A.V.F., D.S.K. and N.V.U.; Formal analysis and investigation: D.I.F., I.S.V., A.A.O. and A.V.F.; Writing—original draft preparation: D.I.F. and I.S.V.; Writing—review and editing D.S.K. and N.V.U.; Funding acquisition: N.V.U.; Resources: I.S.V., D.S.K. and N.V.U.; Supervision: D.S.K. and N.V.U. All authors have read and agreed to the published version of the manuscript.

Funding: This research was funded by the Russian Science Foundation, grant No. 22-13-20071.

Institutional Review Board Statement: Not applicable.

Informed Consent Statement: Not applicable.

Data Availability Statement: The data presented in this study are available in the article and Supplementary Materials.

Acknowledgments: This study was performed using the instrumentation of the Core Facility Center "Arktika" of the Lomonosov Northern (Arctic) Federal University.

Conflicts of Interest: The authors declare no conflict of interest.

References

1. MacRae, W.D.; Towers, G.H.N. Biological activities of lignans. *Phytochemistry* **1984**, *23*, 1207–1220. [CrossRef]
2. Saleem, M.; Hyoung, J.K.; Ali, M.S.; Yong, S.L. An update on bioactive plant lignans. *Nat. Prod. Rep.* **2005**, *22*, 696–716. [CrossRef] [PubMed]
3. Calvo-Flores, F.G.; Dobado, J.A.; Isac-García, J.; Martín-MartíNez, F.J. *Lignin and Lignans as Renewable Raw Materials: Chemistry, Technology and Applications*; John Wiley & Sons: Chichester, UK, 2015; pp. 313–454. [CrossRef]
4. Willför, S.M.; Ahotupa, M.O.; Hemming, J.E.; Reunanen, M.H.T.; Eklund, P.C.; Sjöholm, R.E.; Eckerman, C.S.E.; Pohjamo, S.P.; Holmbom, B.R. Antioxidant Activity of Knotwood Extractives and Phenolic Compounds of Selected Tree Species. *J. Agric. Food Chem.* **2003**, *51*, 7600–7606. [CrossRef] [PubMed]
5. Kumar, R.; Tsvetkov, D.E.; Varshney, V.K.; Nifantiev, N.E. Chemical constituents from temperate and subtropical trees with reference to knotwood. *Ind. Crop. Prod.* **2020**, *145*, 112077. [CrossRef]
6. Ul'yanovskii, N.V.; Onuchina, A.A.; Faleva, A.V.; Gorbova, N.S.; Kosyakov, D.S. Comprehensive Characterization of Chemical Composition and Antioxidant Activity of Lignan-Rich Coniferous Knotwood Extractives. *Antioxidants* **2022**, *11*, 2338. [CrossRef] [PubMed]
7. Willfor, S.M.; Smeds, A.I.; Holmbom, B.R. Chromatographic analysis of lignans. *J. Chromatogr. A* **2006**, *1112*, 64–77. [CrossRef] [PubMed]
8. Kuehnl, S.; Schroecksnadel, S.; Temml, V.; Gostner, J.M.; Schennach, H.; Schuster, D.; Schwaiger, S.; Rollinger, J.M.; Fuchs, D.; Stuppner, H. Lignans from *Carthamus tinctorius* suppress tryptophan breakdown via indoleamine 2,3-dioxygenase. *Phytomedicine* **2013**, *20*, 1190–1195. [CrossRef] [PubMed]
9. Ekiert, R.J.; Szopa, A.; Ekiert, H.; Krzek, J.; Dzik, E. Analysis of lignans in *Schisandra chinensis* fruits, leaves, biomasses from in vitro cultures and food supplements. *J. Funct. Foods* **2013**, *5*, 1576–1581. [CrossRef]
10. Srivastava, V.; Singh, M.; Malasoni, R.; Shanker, K.; Verma, R.K.; Gupta, M.M.; Gupta, A.K.; Khanuja, S.P.S. Separation and quantification of lignans in Phyllanthus species by a simple chiral densitometric method. *J. Sep. Sci.* **2008**, *31*, 47–55. [CrossRef]
11. Sukumar, D.; Arimboor, R.; Arumughan, C. HPTLC fingerprinting and quantification of lignans as markers in sesame oil and its polyherbal formulations. *J. Pharm. Biomed. Anal.* **2008**, *47*, 795–801. [CrossRef]
12. Sarajlija, H.; Čukelj Mustač, N.; Novotni, D.; Mršić, G.; Brncic, M.; Curic, D. Preparation of Flaxseed for Lignan Determination by Gas Chromatography-Mass Spectrometry Method. *Czech J. Food Sci.* **2012**, *30*, 45. [CrossRef]

13. Bonzanini, F.; Bruni, R.; Palla, G.; Serlataite, N.; Caligiani, A. Identification and distribution of lignans in *Punica granatum* L. fruit endocarp, pulp, seeds, wood knots and commercial juices by GC–MS. *Food Chem.* **2009**, *117*, 745–749. [CrossRef]
14. Kebbi-Benkeder, Z.; Colin, F.; Dumarçay, S.; Gérardin, P. Quantification and characterization of knotwood extractives of 12 European softwood and hardwood species. *Ann. For. Sci.* **2015**, *72*, 277–284. [CrossRef]
15. Anikeenko, E.A.; Ul'yanovskii, N.V.; Shavrina, I.S.; Kosyakov, D.S. Laser Desorption/Ionization of Low-Molecular-Weight Lignin Oligomers. *J. Anal. Chem.* **2020**, *75*, 1814–1824. [CrossRef]
16. Sicilia, T.; Niemeyer, H.B.; Honig, D.M.; Metzler, M. Identification and stereochemical characterization of lignans in flaxseed and pumpkin seeds. *J. Agric. Food Chem.* **2003**, *51*, 1181–1188. [CrossRef] [PubMed]
17. Meagher, L.P.; Beecher, G.R.; Flanagan, V.P.; Li, B.W. Isolation and characterization of the lignans, isolariciresinol and pinoresinol, in flaxseed meal. *J. Agric. Food Chem.* **1999**, *47*, 3173–3180. [CrossRef] [PubMed]
18. Avula, B.; Choi, Y.W.; Srinivas, P.V.; Khan, I.A. Quantitative determination of lignan constituents from *Schisandra chinensis* by liquid chromatography. *Chromatographia* **2005**, *61*, 515–518. [CrossRef]
19. Gu, W.; Wei, N.; Wang, Z. LC analysis of lignans from *Schisandras sphenanthera* Rehd. et Wils. *Chromatographia* **2008**, *67*, 979–983. [CrossRef]
20. Milder, I.E.J.; Arts, I.C.W.; Venema, D.P.; Lasaroms, J.J.P.; Wähälä, K.; Hollman, P.C.H. Optimization of a liquid chromatoraphy-tandem mass spectrometry method for quantification of the plant lignans secoisolariciresinol, matairesinol, lariciresinol, and pinoresinol in foods. *J. Agric. Food Chem.* **2004**, *52*, 4643–4651. [CrossRef]
21. Popova, I.E.; Hall, C.; Kubatova, A. Determination of lignans in flaxseed using liquid chromatography with time-of-flight mass spectrometry. *J. Chromatogr. A* **2009**, *1216*, 217–229. [CrossRef]
22. Cacciola, F.; Rigano, F.; Dugo, P.; Mondello, L. Comprehensive two-dimensional liquid chromatography as a powerful tool for the analysis of food and food products. *Trends Anal. Chem.* **2020**, *127*, 115894. [CrossRef]
23. Stoll, D.R.; Li, X.; Wang, X.; Carr, P.W.; Porter, S.E.G.; Rutan, S.C. Fast, comprehensive two-dimensional liquid chromatography. *J. Chromatogr. A* **2007**, *1168*, 3–43. [CrossRef] [PubMed]
24. Cacciola, F.; Jandera, P.; Mondello, L. Comparison of high-temperature gradient heart-cutting and comprehensive LCxLC systems for the separation of phenolic antioxidants. *Chromatographia* **2007**, *66*, 661–667. [CrossRef]
25. Chen, X.; Jiang, Z.; Zhu, Y.; Tan, J. Use of two-dimensional liquid chromatography combined with diode-array and mass spectrometric detection for analysis of compounds in *Flos Lonicera*. *Chromatographia* **2007**, *65*, 141–147. [CrossRef]
26. Wang, S.; Wang, C.; Zhao, X.; Mao, S.; Wu, Y.; Fan, G. Comprehensive two-dimensional high performance liquid chromatography system with immobilized liposome chromatography column and monolithic column for separation of the traditional Chinese medicine *Schisandra chinensis*. *Anal. Chim. Acta* **2012**, *713*, 121–129. [CrossRef] [PubMed]
27. Camenzuli, M.; Schoenmakers, P.J. A new measure of orthogonality for multi-dimensional chromatography. *Anal. Chim. Acta* **2014**, *838*, 93–101. [CrossRef] [PubMed]
28. Gilar, M.; Olivova, P.; Daly, A.E.; Gebler, J.C. Orthogonality of Separation in Two-Dimensional Liquid Chromatography. *Anal. Chem.* **2005**, *77*, 6426–6434. [CrossRef] [PubMed]
29. Lazzari, E.; Arena, K.; Caramão, E.B.; Mondello, L.; Herrero, M. Comprehensive two-dimensional liquid chromatography-based quali-quantitative screening of aqueous phases from pyrolysis bio-oils. *Electrophoresis* **2021**, *42*, 58–67. [CrossRef]
30. Willför, S.; Hemming, J.; Reunanen, M.; Holmbom, B. Phenolic and lipophilic extractives in Scots pine knots and stemwood. *Holzforschung* **2003**, *57*, 359–372. [CrossRef]
31. Jomová, K.; Hudecova, L.; Lauro, P.; Simunkova, M.; Alwasel, S.H.; Alhazza, I.M.; Valko, M. A Switch between Antioxidant and Prooxidant Properties of the Phenolic Compounds Myricetin, Morin, 3′,4′-Dihydroxyflavone, Taxifolin and 4-Hydroxy-Coumarin in the Presence of Copper (II) Ions: A Spectroscopic, Absorption Titration and DNA Damage Study. *Molecules* **2019**, *24*, 4335. [CrossRef]
32. Alperth, F.; Schneebauer, A.; Kunert, O.; Bucar, F. Phytochemical Analysis of Pinus cembra Heartwood—UHPLC-DAD-ESI-MSn with Focus on Flavonoids, Stilbenes, Bibenzyls and Improved HPLC Separation. *Plants* **2023**, *12*, 3388. [CrossRef]
33. Bede, J.C.; Tobe, S.S. Insect juvenile hormones in plants. *Stud. Nat. Prod. Chem.* **2000**, *22*, 369–418. [CrossRef]
34. Ul'yanovskii, N.V.; Onuchina, A.A.; Ovchinnikov, D.V.; Faleva, A.V.; Gorbova, N.S.; Kosyakov, D.S. Analytical and Preparative Separation of Softwood Lignans by Supercritical Fluid Chromatography. *Separations* **2023**, *10*, 449. [CrossRef]

molecules

Article

Development, Validation, and Two-Year Application of Rapid and Simple LC-MS/MS-Based Method for the Determination of K2MK-7 in Blood Samples

Łukasz Paprotny [1], Dorota Szewczak [1], Iryna Bryshten [2] and Dorota Wianowska [2,*]

[1] ALAB Laboratories, Research and Development Centre, ul. Ceramiczna 1, 20-150 Lublin, Poland; dorota.szewczak@alab.com.pl (D.S.)

[2] Department of Chromatography, Institute of Chemical Sciences, Faculty of Chemistry, Maria Curie-Skłodowska University in Lublin, Pl. Maria Curie-Skłodowska 3, 20-031 Lublin, Poland

* Correspondence: dorota.wianowska@mail.umcs.pl

Abstract: Biological properties of menaquinone-7, one of the vitamin K2 vitamers (K2MK-7), both those proven and those that remain to be investigated, arouse extensive interest that goes beyond the strictly scientific framework. The most important of them is the prevention of age-related diseases, considering that we live in the times identified as the era of aging societies and many people are exposed to the vitamin K2MK-7 deficiency. Therefore, an effective analytical protocol that can be adopted as a diagnostic and preventive analytics tool is needed. Herein, a simple sample preparation method followed by the liquid chromatography-tandem mass spectrometry-based method (LC-MS/MS), was used for the selective and sensitive determination of K2MK-7 in serum samples. Under the optimized conditions, using 500 μL of serum and the same amount of *n*-hexane, the reproducibility and the accuracy were obtained in the ranges of 89–97% and 86–110%, respectively, and the limit of detection value was 0.01 ng/mL. This method was used for the routine analysis. Statistical interpretation of the data from 518 samples obtained during 2 years of practice allowed for obtaining information on the content and distribution of K2MK-7 in the Polish population, broken down by the sex and age groups.

Keywords: menaquinone-7 analysis; vitamin K vitamers; sample preparation; extraction; chromatographic analysis; diagnostic tool; population variability

1. Introduction

Vitamin K represents a family of structurally related compounds containing the 2-methyl-1,4-naphthoquinone group substituted with different hydrocarbon side chains at the C3 position. Vitamin K with a long phytyl side chain is called K1 (phylloquinone), while that with a long polypropenyl side chain is called K2 and constitutes a group of homologues known as menaquinones-n (MK-n), where "*n*" stands for a number of isoprenoid units in the side chain. Both forms are found in nature, but as K1 is found mainly in green plants and algae, K2 is synthesized by bacteria and can be found in both animal products such as meat, cheese or fermented food products, and in the human digestive tract [1]. Owing to the presence of the naphthoquinone ring, both forms exhibit numerous and more or less specific biological activities, the impact of which on the human body was underestimated not long ago [2,3]. However, only the results published at the beginning of the 21st century, proving the relationship between senile diseases and vitamin K deficiency, fully reveal its importance [2,4]. Taking into account the main health problems of the modern world and the fact that aging of the population is one of the dominant trends in the 21st century, the most important biological properties are the prevention of osteoporosis and cardiovascular diseases as well as the reduction in the risk of cancer [3–7]. According to [2–6,8] these properties are more specific to MK-7, hence the growing interest in determining this compound is observed. This interest goes beyond the strictly scientific framework, and currently, the analysis of vitamin K2MK-7 becomes an important diagnostic and preventive tool [9]. This is because the half-life of MK-7 is several days, not 1–2 h as in the case of K1, and the longer half-life means that it stays longer in the blood and is much more available to extrahepatic tissues [10]. In addition, there is a large difference in the steady-state concentration of the two forms of the vitamin, and the serum concentration of MK-7 is significantly higher than the concentration of vitamin K1 when equimolar amounts are administered to patients [11]. Considering the wide area of action of the vitamin and, among others, bioavailability issues, many people are at risk of vitamin K2MK-7 deficiency. This applies especially to people with atherosclerosis, hypercholestrolemia, diabetes, dysbiosis, celiac disease, inflammatory bowel diseases, diseases of the liver and biliary tract, people undergoing long-term oral antibiotic therapy, as well as the elderly and obese ones [1,12,13]. Allied to this, it is necessary to develop a simple analytical protocol useful in the laboratory practice that will allow for selective and sensitive monitoring of K2MK-7 levels in biofluids and at the same time accelerating sample preparation and obtaining higher laboratory throughput. This task is not easy due to low dietary intake and consequently, the low concentration of this compound in the human bloodstream [1,14].

Several techniques are applied for vitamin K determination in the biological samples. These are both indirect and direct approaches. Indirect methods, an example of which may be the measurement of one of the biochemical indicators, i.e., γ-carboxyglutamic acid in urine, allow us to assess whether the body lacks vitamin K and in no way measure the level of vitamin K, let alone the level of individual vitamin K [15]. In direct methods, migration separation techniques are used for the determination of K2MK-7. Among them, liquid chromatography (LC) with the fluorescent detection after the post-column zinc reduction to a stable hydroquinone derivative, is the most commonly applied [12,16,17]. The most recent one is LC coupled with mass spectrometry (MS) or tandem mass spectrometry (MS/MS) detection [9,15,18–20]. These techniques have great analytical capabilities; however, are insufficient to determine a compound at low concentration levels in a complex natural matrix [21–24]. Proper sample preparation is still needed, which will not only enable the detection of the non-fluorescent analyte, but also allow for the complete isolation of the compound from bioliquids and its concentration.

Most approaches for the determination of K2MK-7 use solid-phase extraction (SPE) and/or liquid–liquid extraction (LLE), with the most commonly used extractants being those with higher partition coefficients (*log P*), such as diethyl ether (0.9), dichloromethane (1.5), ethyl acetate (0.7), toluene (2.7), chloroform (2.3), *n*-hexane (3.9), and cyclohexane (3.4) [12,17–20,25,26]. These approaches use relatively large volumes of solvents (4–8 mL) in a single extraction step, with several steps usually being used to increase extraction efficiency [9]. These are also time-consuming procedures, usually taking several dozen minutes, which significantly reduces the throughput of the laboratory. In the literature, there are also procedures that require even 4 h to prepare a single sample [15], the implementation of which, under the conditions of laboratory diagnostics, focused on the analysis of a large number of samples per working shift, and speeding up work is difficult if not realistic. For these reasons, it is necessary to develop a quick, simple, cheap, and efficient sample preparation method that can be conducted using equipment typical of diagnostic laboratories. Another important issue is to use the smallest possible amount of biological sample needed for extraction. In addition, the principles of green analytical chemistry emphasized the need to reduce the consumption of toxic and expensive organic solvents. Hence, currently the attention is paid to the use of extraction techniques consuming minimal amounts of organic solvents. In this respect, liquid–liquid microextraction (LLME) is a modern alternative to the conventional LLE extraction technique [27].

Typically, LLME is used in the analysis of water, food, and natural product samples [25]. Despite the great application potential, this technique never received sufficient attention so far in the biomedical analysis. Therefore, the aim of this study is to optimize the extraction procedure for the efficient and selective K2MK7 isolation from the serum samples, in line with modern requirements, and its subsequent quantitative analysis using the recently developed and validated rapid LC-MS/MS-based method. As the optimized method proved to be an effective and useful diagnostic tool, it was used in our laboratory (Research and Development Centre, ALAB, Lublin, Poland) for the routine analyses for a two-year period to obtain information on the content and distribution of K2MK-7 in the Polish population by different gender and different age groups. Data were created from 518 serum samples; 52.1% of the samples were female ones and 47.9% of the male samples. This information will be useful for the subsequent development of K2MK-7 reference ranges in Poland.

2. Results and Discussion

2.1. Optimization of the Sample Preparation Procedure

In this study, in order to optimize extraction procedure for the quantitative analysis of vitamin K2MK-7 in serum, 16 extraction systems were tested, differing in the type of the applied extraction solvent: dichloromethane (CH_2Cl_2), chloroform ($CHCl_3$), diethyl ether (Et_2O), carbon tetrachloride (CCl_4), and *n*-hexane, its volume (250 or 500 μL), and the addition of a precipitant: acetonitrile (ACN), ethyl alcohol (EtOH), and ammonium acetate (AA). Their characteristics are presented in the Section 3. The tests were carried out using the optimized LC-MS method (see the Section 3) and discussed in the Section 2.2. The studies were carried out in the 2 mL Eppendorf vials, using 500 μL of the serum free from the test substance, fortified with the known amount of the vitamin K2MK-7 standard and K2MK-7-D7 at a concentration of 2 ng/mL. The results are presented in Figure 1 as the mean values of the peak areas of the analyte and the internal standard obtained for a given system and calculated from three independent extractions, indicating both the solvent and the factor increasing the extraction efficiency for the sake of clarity discussion. In order to assess the effect of changing the extraction conditions on the analyte peak area, the obtained data were statistically processed using the one-way analysis of variance (ANOVA), and its results are presented in Table 1. For clarity, if the calculated value of F (F_{cal}) exceeds the table value F (F_{tab}), this indicates a statistically significant influence of the given parameter.

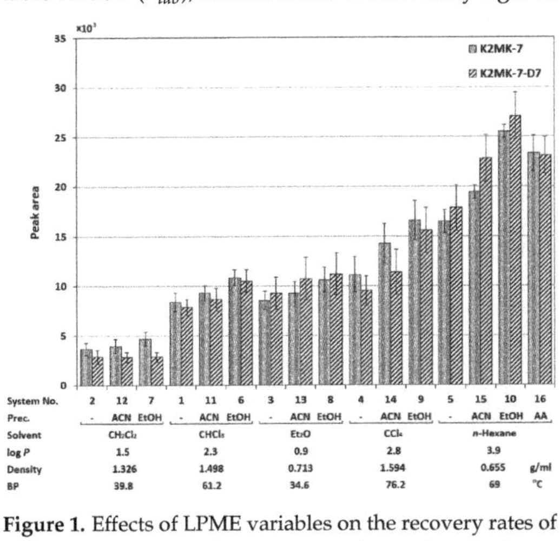

Figure 1. Effects of LPME variables on the recovery rates of K2MK-7 and K2MK-7-D7. Explanation of abbreviations—see the relevant section of the text above.

As can be seen from the data presented in Figure 1, the individual systems are characterized by different isolation efficiencies of the vitamin K2MK-7 standard and its deuterated derivative. The change in the extraction solvent type differentiates clearly the size of the surface area of both compounds (F_{cal} = 48.45, F_{tab} = 3.48, see Table 1). In general, the smallest peak areas were observed in the extracts obtained with CH_2Cl_2 and the largest with *n*-hexane. At the first glance, it might seem that the observed differences are the result of different affinities of individual solvents for the analyte. However, the problem is more complex.

Table 1. Summary of the statistical analysis of the obtained results.

Effects		F_{cal}-Value	p-Value	F_{tab}-Value
Effect of solvent type on K2MK-7 peak area		48.45	1.64×10^{-6}	3.48
Effect of EtOH adding to the extraction system on K2MK-7 peak area		129.85	1.42×10^{-8}	3.48
Effect of ACN adding to the extraction system on K2MK-7 peak area		72.18	2.45×10^{-7}	3.48
Effect of AA adding to the extraction system on K2MK-7 peak area		3.93	0.12	7.71
Effect of CH_2Cl_2 volume		1.82	0.24	5.14
Effect of $CHCl_3$ volume		6.48	0.03	5.14
Effect of Et_2O volume		2.57	0.16	5.14
Effect of CCl_4 volume		6.01	0.04	5.14
Effect of n-hexane volume		83.97	4.10×10^{-5}	5.14
Effect of age on K2MK7 concentration in female		4.18	2.18×10^{-4}	2.05
Effect of age on K2MK7 concentration in male		3.85	5.45×10^{-4}	2.05
Effect of sex on K2MK-7 concentration in individual age-groups:	<10	0.19	0.66	4.26
	11–20	1.26	0.27	4.06
	21–30	0.24	0.63	3.96
	31–40	1.25	0.26	3.92
	41–50	3.97	4.8×10^{-2}	3.94
	51–60	0.14	0.71	3.96
	61–70	2.71	0.10	4.04
	>71	0.06	0.81	4.41
		rho-value	p-value	
Correlation between age and K2MK-7 concentration in female		0.23	2.34×10^{-4}	
Correlation between age and K2MK-7 concentration in male		0.27	2.68×10^{-5}	
Resultant correlation between age and K2MK7 concentration		0.27	2.46×10^{-9}	

If the *log P* value is taken as the measure of affinity, the extraction efficiency should increase with the increase in this value. Meanwhile, the signals of both compounds observed for Et_2O (*log P* = 0.9) are about twice higher compared to those observed for CH_2Cl_2 (*log P* = 1.5). Taking into account the physicochemical properties of these solvents, the possible reason for the observed differences is the lower density of Et_2O than CH_2Cl_2 (see Figure 1). This fact makes it easier to collect the upper lighter ether layer manually. On this basis, it can be concluded that not only the type of solvent (its hydrophobicity), but also the method of the extraction, i.e., the ease of manipulation of the readily accessible upper layer of the extraction solvent, affects the resultant effectiveness of the compounds isolation in the miniaturized extraction systems.

Comparing the effects caused by the addition of the precipitation reagent into the extraction system (which in Figure 1 is designated as "Prec."), it can be seen that the signals of both compounds are generally greater than those obtained previously. This is especially evident for systems with n-hexane and CCl_4. This observation can be explained by achieving greater selectivity of extraction, reducing the background effect in the evaluation of the MS signals magnitude. The effect is greater for EtOH and smaller for can, which is confirmed by the values of the Fischer coefficient from Table 1, equal to 129.85 and 72.18, respectively (F_{tab} = 3.48). However, this result contradicts the information available in the literature that ACN is a more effective precipitating agent compared to ethanol [27]. Nevertheless, according to the literature, ACN can have a negative effect on the formation of ions and can cause the so-called ion suppression phenomenon, leading to a decrease in the measured signals [28,29]. This comment applies not only to the effect of acetonitrile, but also to the other solvents used in the stage of sample preparation for analysis. The truthfulness of this statement seems to confirm the data obtained for ammonium acetate (AA). As it is commonly known, this compound increases the ionization efficiency of analytes in MS [30]. Admittedly, its use in these studies did not increase the signal magnitude of both compounds because there are no statistically significant differences in their peak intensities compared to those observed with the addition of ethanol (F_{cal} = 3.93, F_{tab} = 7.71, see Table 1), but it improved the precision of the signal evaluation significantly (compare the size of the bars in Figure 1).

As for the effect of the volume of the used extractant on the analyte signal, statistically significant differences (see Table 1) were observed only for the *n*-hexane extract, i.e., the solvent showing the largest peak areas of both K2MK-7 and IS. For this extractant, higher signals were observed for a smaller volume of *n*-hexane, i.e., 250 μL. Nevertheless, taking into account the complexity of the biological sample matrix and the assumed low vitamin content in real samples, the use of a smaller volume of the solvent in a single extraction cycle may result in deterioration in the measurement accuracy. The conducted studies on the impact of the number of extraction cycles on the effectiveness of K2MK-7 isolation using 500 and 250 μL of *n*-hexane (data are shown in the Supplementary Materials), showing the need for three extraction cycles for a smaller volume of extractant to achieve comparable accuracy, confirmed the validity of this reasoning. Therefore, it was decided to use 500 μL of *n*-hexane and the same amount of biological material for the final determination. There are also economic reasons behind this choice, as the use of a larger amount of solvent in a single extraction cycle allows for significant speeding up of the sample preparation and for obtaining larger throughput of the laboratory.

2.2. Method Validation

The proposed analytical protocol for the analysis of K2MK-7 in the human serum was successfully validated following the criteria of FDA for the bioanalytical method validation (see the Section 3) [31].

Identification and quantification were based on the MS/MS multiple reaction monitoring (MRM) mode after studying the fragmentation spectra of the analyte and IS in accordance with the confirmation criteria taken from Commission Decision 2002/657/EC. According to them, the MRM measurements for the analytes were performed for two transitions: m/z + 649.5 → 121.0 (qualifier transition, S2) and 649.5 → 187.2 (quantifier transition, S1) for K2MK-7, and one transition m/z + 656.0 → 194.1 for K2MK-7-D7 using the collision energy (CE) values at 41 eV, 30 eV, and 37 eV, respectively (the dwell time of 500 ms). During the optimization experiments, the suitability of the ESI and APCI ionization sources was checked. Although both gave the same fragmentation ions, the intensity of the signals was higher using the APCI source, which determined the use of this ionization source. Figure 2 shows the representative APCI(+)-MRM chromatograms obtained during the validation experiments for the pooled serum samples irradiated with UV light to destroy endogenous vitamin as well as the authentic serum sample, confirming acceptable analytical conditions for both qualitative and quantitative analyses of K2MK-7 in the serum samples. Figure 2A shows the chromatograms of the blank serum sample (upper chromatogram) spiked with deuterium labeled K2MK-7 (K2MK-7-D7) used as the internal standard (lower chromatogram, IS). Figure 2B,C shows the chromatograms obtained for the pooled serum samples with the addition of IS and the analyte at the level of 0.35 ng/mL and 0.96 ng/mL, respectively. Figure 2D shows the chromatograms obtained for an authentic serum sample tested for the content of vitamin K2MK-7. The chemical structure and possible fragmentation pattern of K2MK-7, together with the chemical structure of the product ions used for the analysis, are presented in Figure 3.

The specificity of the method was confirmed by matching the retention times (RT) and the fragmentation ion signal ratio (S2/S1) obtained for the analyte in five different aliquots of the authentic serum samples to the mean values obtained for the calibration solutions (RT_{ref} = 4.01 min ± 2.0% and $S2/S1_{ref}$ = 0.439) with the differences in the retention times at the −0.069 level, while the relative signal difference was −1.82%.

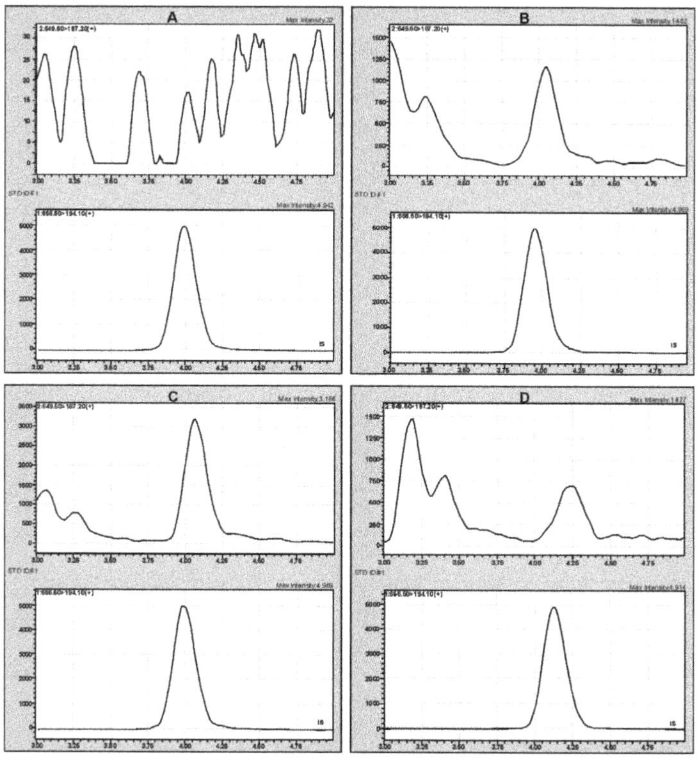

Figure 2. MRM chromatograms of the blank serum sample (**A**), the control sample obtained by spiking the blank serum sample with the analyte at the level of 0.35 ng/mL (**B**) and 0.96 ng/mL (**C**), and the chromatogram of the authentic serum sample tested for the content of vitamin K2MK-7 (**D**).

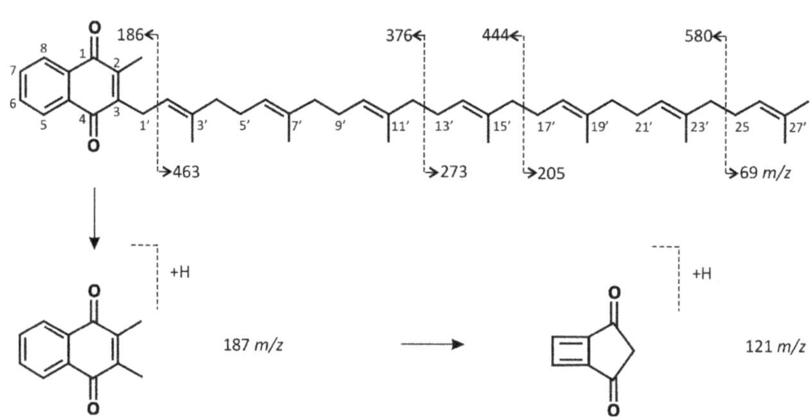

Figure 3. Chemical structure and a characteristic fragmentation pattern of K2MK-7 together with the chemical structure of the product ions used for the analysis.

The calibration curve was constructed not using the weighted model, which resulted in small and evenly distributed residual errors. The curve is shown in in the Supplementary Materials. Its characteristic features determined by the slope, the intercept (±confidence interval at the confidence level of 0.95), and coefficient of determination (R^2)

are 1.30621 (±0.106), −0.00789 (±0.0905), and 0.991, respectively. The F-value obtained for the statistical method of the lack of fit was equal to 0.79 (F_{cal}), with the values of the numerator and the denominator 7 and 18, respectively, and the tabular F-value equal to 2.58 (F_{tab}). $F_{cal} < F_{tab}$ confirms the linear relationship between the ratio of K2MK-7 to IS signals and the K2MK-7 concentration. The back-calculation of K2MK-7 concentrations in the calibration solutions, performed to confirm suitability of the used calibration, showed acceptable deviation of the interpolated concentrations of standards. Inaccuracy and imprecision of K2MK-7 determinations, expressed as % bias and % CV, were in the range of −11.0–5% and 2.3–14.2% for six out of seven non-zero calibration solutions. Only in the case of the smallest standard concentration was the deviation at 18% obtained, and this concentration was accepted as the LLOQ level.

Determination of the signal-to noise ratio performed using the LLOQ sample to establish the smallest analyte content allowing to confirm its presence (LOD) and to quantify it (LOQ) reached 0.013 and 0.039 ng/mL.

The data on the inaccuracy and imprecision of the present study are presented in Table 2. The intra-and inter-day inaccuracy, expressed in % bias, ranged from −13.8 to 10.1% with the precision of 3.3–11% (% CV). These results show adequate great precision and accuracy. The correctness of this statement confirms the ANOVA and the Student's t-test results showing that there is a lack of a significant difference between the results at the individual concentration levels ($1.00 < F > 3.22$, $F_{tab} = 3.48$; $-0.06 < t_{cal} < 2.46$, $t_{crit} = 4.30$). The dilution study carried out at 1:1, 1:8, and 1:16 for the samples fortified above the highest concentration of the calibration standard showed that the dilution can be undertaken with a sufficiently good precision and accuracy. The precision of the study ranged from 6.4 to 12.8% CV and the accuracy ranged between −10.5 and 8.2% bias.

Table 2. Intra- and inter-day precision, accuracy, and summary of stability study for the determination of K2MK-7 in the serum samples.

Nominal Concentration (ng/mL)	Measured Concentration (mean ± SD), (ng/mL)	Imprecision (% CV)	Inaccuracy (% BIAS)
Intra-day (n = 5)			
0.10	0.116 ± 0.004	3.98	16.16
0.32	0.352 ± 0.022	6.40	10.10
0.64	0.728 ± 0.069	9.49	−13.80
0.96	0.939 ± 0.030	3.29	−2.12
Inter-day (n = 5)			
0.10	0.108 ± 0.005	4.51	8.63
0.32	0.317 ± 0.030	9.48	0.98
0.64	0.649 ± 0.072	11.04	−1.38
0.96	1.007 ± 0.073	7.22	−4.88
Autosampler at 15 °C (24 h) (n = 3)			
0.32	0.352 ± 0.01	2.8	−10.0
0.64	0.621 ± 0.084	13.52	3.06
0.96	1.031 ± 0.052	5.04	−12.81
Refrigerator at 4 °C (24 h) (n = 3)			
0.32	0.360 ± 0.06	13.76	12.50
0.64	0.622 ± 0.06	8.59	2.82
0.96	1.014 ± 0.046	4.44	−5.62
Freezer at −18 °C (24 h) (n = 3)			
0.32	0.325 ± 0.042	12.92	1.56
0.64	0.619 ± 0.033	5.33	3.28
0.96	1.014 ± 0.046	5.80	6.66

Table 2. *Cont.*

Nominal Concentration (ng/mL)	Measured Concentration (mean ± SD), (ng/mL)	Imprecision (% CV)	Inaccuracy (% BIAS)
Exposed to light at room temp. (1 h) (n = 3)			
0.32	0.167 ± 0.004	2.39	47.81
0.64	0.250 ± 0.188	75.08	60.88
0.96	0.344 ± 0.240	69.79	64.18
Exposed to light at room temp. (2 h) (n = 3)			
0.32	0.125 ± 0.060	47.83	60.80
0.64	0.299 ± 0.092	30.91	53.19
0.96	0.508 ± 0.12	23.60	47.03
Freeze-thaw cycles (cycle number) (n = 3)			
0.64 (1)	0.600 ± 0.014	2.33	6.20
0.64 (2)	0.716 ± 0.041	5.74	−11.94
0.64 (3)	0.657 ± 0.058	8.83	−2.65
0.64 (4)	0.379 ± 0.052	13.71	40.66

Various storage and handling conditions were evaluated to determine their effect on the vitamin K2MK-7 stability. The obtained results are given in Table 2. The samples kept in darkness in the autosampler, refrigerator, and freezer at a temperature from 15 to −18 °C for 24 h are stable. Alternatively, the samples exposed to light at room temperature show a loss exceeding the accepted stability criterion after just one hour [25,32]. The stability study after four freeze–thaw cycles showed that the analyte is stable up to three cycles. These data prove that, while the low temperature under the tested conditions stabilizes K2MK-7, one should avoid definitely exposing the sample to UV radiation. Moreover, multiple freezing and thawing of serum samples subjected to the K2MK-7 determination should be avoided.

Table 3 summarizes the results of the matrix effect (ME), recovery (RE), and process efficiency (PE) studies (for details, see the Section 3). These parameters were obtained using the method described in [33] by determining the area of the K2MK-7 and IS peaks for three differently prepared sets of samples, each at three concentration levels. The parameters are calculated as a percentage of the response of set 2 samples in relation to those of set 1 samples (ME), the response of set 3 samples in relation to that of set 2 (RE), and the response of set 3 samples in relation to that of set 1 samples (PE) [33]. The obtained values confirm a slight increase in the ionization of the analyte by co-eluting substances from the biological matrix. This effect is especially pronounced at smaller analyte concentrations. Nevertheless, the efficiency of the process is comparable and high. The recovery ranges from 79 to 85% and from 84 to 88% for the analyte and IS, respectively. These values are suitable for the quantitative K2MK-7 analysis in the human serum.

Table 3. Summary of matrix effect (ME), recovery (RE), and process efficiency (PE) studies for K2MK-7 and IS at three levels of vitamin concentrations covering the calibration range for three differently prepared sets of samples: set 1 is samples of standard solutions prepared in water instead of serum, sets 2 and 3 are pooled blank serum samples that were fortified with standard solutions after extraction (set 2) and before extraction (set 3).

Nominal Concentration (ng/mL)	Mean Peak Area (% CV)						ME (%)		RE (%)		PE (%)	
	K2MK-7			IS								
	Set 1	Set 2	Set 3	Set 1	Set 2	Set 3	K2MK-7	IS	K2MK-7	IS	K2MK-7	IS
0.32	16,954 (5.21)	19,281 (1.34)	15,280 (9.54)	35,768 (2.55)	38,317 (2.25)	32,313 (9.14)	113.73	107.13	79.25	84.33	90.13	90.34
0.64	33,288 (10.87)	36,584 (9.12)	30,011 (9.20)	32,235 (7.76)	36,701 (1.28)	31,833 (15.70)	109.90	113.85	82.03	86.74	90.15	98.75
0.96	48,747 (4.08)	49,094 (0.72)	41,721 (2.58)	33,729 (11.95)	34,329 (2.13)	30,039 (10.51)	100.71	101.78	84.98	87.50	85.59	89.06

2.3. *Practical Application*

To assess the applicability of the optimized method for the routine laboratory practice, studies were carried out in the group of 10 healthy volunteers with normal dietary habits. The research confirmed that the method is suitable for the analysis of vitamin K2MK-7 in the real serum samples. Therefore, after carrying out comparative tests showing compliance of the results at the level of 85–90% with the other laboratory tests (see the Supplementary Materials), the decision was made to implement the developed method as an analysis routinely performed in our laboratory. The results presented below constitute the statistical interpretation of the data obtained during the standard practice of our laboratory during the 2-year period. As mentioned, these data were obtained from 518 serum samples; 52.1% of the samples were female ones and 47.9% were male samples. The K2MK-7 levels in the individual age ranges: <10, 11–20, 21–30, 31–40, 41–50, 51–60, 61–70, and >70 were created based on the analysis of the following number of female (male) samples: 14 (12), 20 (26), 42 (37), 60 (54), 54 (37), 38 (53), 32 (19), and 10 (10). These data are important because, to our knowledge, these are the first available literature data on the content of vitamin K2MK-7 in the Polish community. As a result of the life expectancy extension, as well as changes in the age structure of the population, the median age of Poles as well as of other nationalities is growing year by year. According to WoldData.info [34], among 119 countries assessed in 2018–2020, Poland ranks 28th with the average age of 41.9, which means that half of the Polish population is older than 41.9. Japan is in the lead in this ranking, with the average age of 48.6 years. The USA, for comparison, ranks 44th with the average age of 38.5 years.

The graphical representation of the obtained data is shown in Figure 4 in the form of box plots with the division into the sex- and age-related groups. Their analysis allows us to conclude that, in line with the expectations, the content of vitamin K in the Polish society is generally small. However, as can be seen in the studied age groups for individual genders, the obtained field ranges, their position, median values (marked as the line across the box), and average values (marked with the cross) are different. These features independently confirm the applicability of the developed method for the routine analysis of vitamin K2MK-7 content in the serum samples.

The more detailed analysis of the presented data shows that an asymmetric distribution of values is observed in each group. With age, the width of the box ranges for both genders increases, which means that the degree of results dispersion increases. The highest degree of agreement of the results with the lowest number of outliers and the shortest whiskers is visible for the youngest age group, with the mean and median values in the female and male groups being 0.319 and 0.154 ng/mL as well as 0.387 and 0.239 ng/mL, respectively. In turn, the lowest degree of agreement with the longest whiskers can be observed in the group of males in the middle age group, i.e., 41–50 years old, with the highest values of outliers, not shown in Figure 4 for the sake of legibility, obtained in the oldest age group of females (11.7 ng/mL). In this age group, the highest mean and median values for female and male are also noticeable. These values are 2.14 and 0.73 ng/mL for female and 1.98 and 1.18 ng/mL for male. This observation is consistent with the general knowledge about the increasing incidence of cardiovascular diseases with age and the effect of blood thinning drugs such as warfarin on the content of vitamin K in the body [5,13,35]. Moreover, the smaller content of vitamin K2MK-7 in females compared to its content in males is consistent with the greater frequency of osteoporosis in females [12]. The difference in the amount of vitamin K2MK-7 in females and males is noticeable in the age group 41–50, which is when menopause often begins. These observations, as well as other median values in the figure, visible in the form of a different position of the line in the box, may suggest the probability of a relationship of age and gender with the vitamin K content in the human body. This suggestion was verified by the statistical analysis, the results of which are collected in Table 1, together with those of the statistical analysis of factors influencing the stage of sample preparation. The obtained Spearman's rank correlation coefficients (*rho value*, see Table 1) revealed that the K2MK-7 values are significantly associated with the age both in females ($r = 0.23$, $p < 0.001$) and males (0.27, $p < 0.001$). Higher values of the *F*-test obtained for females (F_{cal} = 4.18, $p < 0.001$) not only confirmed the relationship between the age and gender with the content of vitamin K, but also indicated that in the female group, there is a greater distribution of K2MK-7 values depending on age.

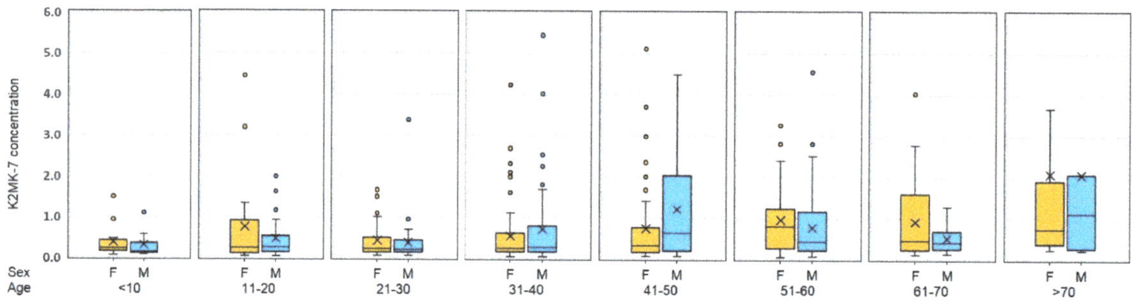

Figure 4. Box plot of the serum concentration of K2MK-7 for different age groups and gender.

2.4. Comparison with the Published Data

Due to the great interest in the role of vitamin K2MK-7 in the human body, and consequently, the sensitive analysis of this compound in the physiological fluids, a number of different approaches to its direct determination can now be found in the literature, including the LC-MS/MS method [12,13,15–18]. The disadvantages of most of these methods are the consumption of large amounts of both sample and other organic solvents, quite complicated equipment used at the sample preparation stage, long analysis time, and high LOD values. Riphagen et al. [19] published the LC-MS method with APCI for determination of vitamins K1, K2MK-4, and K2MK-7 with the simplified pretreatment sample procedure but the limit of quantification for K2MK-7 was only 2.2 ng/mL. This value is definitely not suitable for monitoring the vitamin levels in the healthy population, and even less so for assessing its deficiency. On the other hand, Dunovska et al. [36] proposed the method with the LOD value of 0.002 ng/mL, but with the sample preparation procedure taking up to 4 h, which excludes the possibility of its use in the diagnostic laboratory. Our method definitely stands out because by combining the simplicity of the miniaturized LLE technique with the possibilities of the LC-MS/MS method, we obtained a fast and cheap method that requires a 0.5 mL serum sample and only 0.5 mL extractant to guarantee the high analyte recovery with the small LOD value in about 5 min, with the total analytical run time of 10 min needed to elute the more hydrophobic interfering compounds. Justice should be given, as due to the low content of K2MK-7 in real patient samples, we were unable to reduce the volume of serum needed for extraction in a single extraction run. Nevertheless, although the volume proposed in our methodology is larger than the 350 μL used by Riphagen, it is analogous to the volume used by Dunovska, with a total of 8 mL of solvent used for extraction.

The resulting significant shortening of the overall analysis time, with a radical reduction in the exposure of the unstable analyte to stress factors, made it possible to notice subtle relationships between the age and the gender and the content of vitamin K. These dependencies were not noticed by Dunovska et al. [36], who reported the content of vitamin K2MK-7 in the population of Caucasian as 0.074–0.759 ng/mL for both females and males.

3. Materials and Methods

3.1. Sample Preparation

Methanol (MeOH), ethanol (EtOH), and acetonitrile (ACN), all of LC/MS grade, were purchased from Merck (Darmstadt, Germany). Carbon tetrachloride (CCl_4), chloroform ($CHCl_3$), diethyl ether (Et_2O), methylene chloride (CH_2Cl_2), *n*-hexane, and ammonium acetate (AA) (all with analytical purity grade) were purchased from the Polish Chemicals Plant POCh (Gliwice, Poland), and 99% formic acid of LC-MS purity was from VWR Chemicals (Gdańsk, Poland). The certified standards of K2MK-7 and deuterium-labeled

K2MK-7 (K2MK-7-D7) were obtained from Sigma-Aldrich (St. Louis, MO, USA). The first standard was delivered as a solution in ACN at the concentration of 100 μg/mL. K2MK-7-D7 used as the internal standard (IS) was in the powder form in the batch of 1 mg. It was dissolved in 1.5 mL of EtOH as recommended by the supplier.

Stock solutions of standards were prepared in EtOH by diluting the K2MK-7 standard solution to the concentrations of 10 and 100 ng/mL and the IS standard solution to 50 ng/mL. These solutions were used for calibration, validation, and analysis of real samples. They were kept under the stable conditions at −20 °C (±2 °C) until their use. All other solutions applied for determining validation parameters were prepared on the ongoing basis, unless otherwise stated, from the individual stock standard solutions, being diluted in water and/or in aliquots of the pooled blank serum sample under the conditions protecting them against light.

The quality control (QC) samples were prepared in the pooled serum sample spiked with the individual stock standard solutions. To avoid multiple thawing, these samples were thoroughly mixed to ensure homogeneity, then divided into smaller portions that were separately frozen and stored at −20 °C (±2 °C) until needed.

All samples used in the study were kept in the amber nontransparent vials unless stated otherwise. Water was purified using the Milli-Q system (Millipore Sigma, Bedford, MA, USA).

3.2. *Collection and Storage of Serum Samples*

During the assessment of the suitability of the developed method for the routine laboratory practice, the samples for testing were obtained from 10 healthy volunteers with normal dietary habits after an overnight fasting period. Small volumes of blood samples were collected by the qualified staff in accordance with the local, national, and international regulations (with the Declaration of Helsinki) on ethics after obtaining the informed consent from each of them. Blood was obtained by venipuncture into the 4.9 mL tubes using a single closed system containing the Monovette coagulation activator according to the manufacturer's instructions (Sarstedt AG, Nümbrecht, Germany). Then, it was thoroughly mixed in order to maintain its homogeneity and immediately protected against light using aluminum foil. The serum was separated within 45 min of blood collection by centrifugation (2000× *g*; 10 min at room temp.) and frozen at −20 °C until testing.

Age- and gender-specific K2MK-7 intervals in the Polish population were created using the data obtained from the samples collected during our laboratory standard practices (Research and Development Center, Alab Laboratories, Lublin, Poland). Blood was collected from the patients only during their laboratory tests (no additional material was collected).

3.3. *Sample Preparation*

The optimized procedure consists of mixing 500 μL of the serum sample mechanically with 5 μL of the internal standard solution and 245 μL of EtOH in the 2 mL amber Eppendorf vial, then adding 500 μL of *n*-hexane, 5 min vortexing, and finally centrifugation (14,500 rpm, 2 min). Then, the supernatant upper layer is quantitatively transferred to the dark glass vial, evaporated to dryness under a stream of nitrogen, then reconstituted in 50 μL of methanol with 0.1% formic acid, and after transfer to the insert vials, analyzed.

In order to determine the optimal extraction conditions for the quantitative determination of K2MK-7 in serum, the influence of the following factors on the analyte and IS peak signals was investigated: the type of the extraction solvent, the ratio of volumes of the extraction solvent to the sample, and the addition of a precipitating reagent. The experiments were conducted in 16 systems. Their characteristics are given below.

In systems 1–5 to 500 μL of serum, 500 μL of the suitable solvent were added, i.e., $CHCl_3$ (system 1), CH_2Cl_2 (system 2), Et_2O (3), CCl_4 (4), and *n*-hexane (5). In systems 6–10 to 500 μ of serum and 250 μ of the solvent, i.e., $CHCl_3$ (in system 6), CH_2Cl_2 (7), Et_2O (8), CCl4 (9), n-hexane (10), and also 250 μL of EtOH were added each time. In systems 11–15 to 500 μL of serum and 250 μL of the solvent, i.e., $CHCl_3$ (in system 11) or CH_2Cl_2 (12)

or Et_2O (13) or, CCl_4 (14) or *n*-hexane (15), and 250 µL of ACN were added each time. In system 16, as before, 500 µL of serum and 250 µL of *n*-hexane were used with the addition of 100 µL of 1 M ammonium acetate (AA) solution.

For the system characterized by the highest intensity of analyte and IS signals, the effect of the number of extraction cycles was also examined. The influence of 3 extraction cycles on the K2MK-7 extraction efficiency was taken into account. In all these experiments, a pooled serum was spiked with the analyte and IS at 2 ng/mL. For each of the tested systems, three independent repetitions were made (n = 3).

3.4. LC-APCI-MS Analysis and Its Optimization

The LC analyses were performed using the Shimadzu NEXERA X2 LC system (Shimadzu, Kyoto, Japan) composed of a parallel double binary pump (LC-30AD), a system controller (CBM-20A), an automatic solvent degasser (DGU-20A5R), and an autosampler (SIL-30AC). The autosampler temperature was maintained at 15 °C. Separations were made using the Kinetex C18 column (50 × 2.1 mm, 2.6 µm, Phenomenex, Inc., Torrance, CA, USA) applying gradient elution with the mobile phase flow at 0.6 mL/min. The mobile phase A was the solution of 0.1% formic acid in water; the mobile phase B was 0.1% formic acid in MeOH. The elution conditions were as follows: isocratic at 97% B for 4 min, gradient to 100% B from 4.1 min to 5 min, and isocratic at 100% B to 10 min. At 10.1 min, the mobile phase was reached again 97% B for 10 min to prepare the column for the next sample. The column temperature was maintained at 40 °C and controlled with the column oven (CTO-20AC). The injected sample volume was 30 µL.

Detection was performed with the LCMS-8050 triple quadrupole mass spectrometer (Shimadzu, Kyoto, Japan) equipped with the atmospheric pressure chemical ionization (APCI) source operating in the positive ion mode under the following conditions: the APCI temperature 350 °C, the desolvation line temperature 200 °C, the heating block temperature 200 °C, the nebulizer gas flow 3 L/min, and the drying gas flow 5 L/min.

To determine the optimal conditions for the chromatographic separation, the influence of the mobile phase composition (MeOH and ACN), the volume of formic acid added to the mobile phase (0.05% and 0.1%), the flow rate, the gradient elution profile, the column temperature, and the volume of the injected sample were investigated. To establish the MS/MS operating conditions, the analyte and IS standards, each at the concentration of 100 ng/mL, were separately injected into the LC-MS/MS system. For each compound, mass transitions of the most sensitive or selective precursor ions were optimized in terms of their product ions and corresponding collision energy.

3.5. Method Validation

The validation method was used according to the general validation criteria in terms of specificity, linearity, the limit of quantitation (LOQ), the intra-day and inter-day precision and accuracy, matrix effects, as well as stability of measurements in compliance with the main FDA guidelines for the bioanalytical method validation [31]. In addition, the dilution precision was included as the validation parameter, taking into account a possible higher concentration level of the analyte after its supplementation. Due to the lack of commercially available serum with appropriate characteristics, process optimization, calibration, and validation were performed using a representative portion of serum obtained from the individual volunteers' blood samples that were pooled and checked for the presence of vitamin K2MK-7 after their UV irradiation to photodegrade the endogenous vitamin. After combining, the serum was mixed thoroughly to ensure its homogeneity and divided into smaller portions that were separately frozen. There are no commercially available K2MK-7 quality control (QC) samples; therefore, the aliquots of the blank serum samples were fortified with stock standards solutions to obtain low, medium, and high QC samples. During the experiments, the samples were prepared daily, except for those used to estimate the measurements stability.

To establish a calibration curve, the aliquots of the blank pooled serum sample were spiked with the sequentially increased amounts of the K2MK-7 standard and the same amount of the deuterium-labeled K2MK-7 standard used as the internal standard (IS, 10 ng/mL). Nine calibration samples were examined, including the blank sample, the zero sample (the blank sample with added IS), and seven non-zero samples in the range of 0.1 ng/mL to 1.2 ng/mL of K2MK-7. Three replicated analytical procedures were applied independently for each examined concentration level. A calibration curve was constructed by plotting the peak area ratio of K2MK-7 to IS against the known K2MK-7 concentration. The slope, intercept, and coefficient of determination (R^2) were determined by the least squares linear regression model. To assess the quality of linearity, the statistical approach of lack of correlation was used. The quality of calibration was evaluated by the back-calculation of K2MK-7 concentrations in the calibration solutions. The lowest limit of quantification (LLOQ) was obtained on the basis of the smallest concentration of K2MK-7 that gives the CV and bias values $\leq$ 20%.

The limits of detection and quantification values, LOD and LOQ, respectively, were determined from the analysis of the sample chromatogram obtained for the blank plasma samples enriched with the analyte at the LLOQ level. The LOD and LOQ were considered to be the signal to the noise ratios equal to 3 and 10, respectively.

The intra- and inter-day precision and accuracy were assayed using the QC samples at three concentration levels covering the calibrating range of K2MK-7 concentrations (0.32, 0.64, and 0.96 ng/mL) and the LLOQ level (0.10 ng/mL), and evaluated by the statistical analysis. Five replicates per concentration for the independently prepared samples were analyzed three times on the same day and on five different days within three weeks. The precision was estimated using the one-way analysis of variance (ANOVA) test, and it was expressed as the coefficient of variation (CV, %). The accuracy estimation was made comparing the mean value of the obtained results to the nominal concentration level of the analyte in the control sample using the Student's *t*-test, and it was expressed as BIAS (in %). The precision and accuracy were also tested for the pooled serum samples fortified with the standards above the concentration of the highest calibration standard. In these experiments, the samples were appropriately diluted with the blank serum to give concentrations within the calibration range. The assays were performed preparing the samples at three concentration levels (1.5, 6, and 12 ng/mL) that were diluted to 1:1, 1:8, and 1:16, respectively. The analysis of the diluted samples was performed in the set of 3 replicates for each concentration. Then, the accuracy and precision were determined.

Matrix effect (ME), recovery (RE), and process efficiency (PE) were determined using the method described in [33] by determining the area of the K2MK-7 and IS peaks in three differently prepared sets of samples, each at three concentration levels corresponding to the concentration of the QC samples. The first set was composed of standard solutions prepared in water instead of serum. The second was prepared in the aliquots of the pooled blank sample fortified with the vitamin and IS after the sample preparation. The last one was prepared in the same serum spiked with the standards prior to the extraction.

The stability of K2MK-7 was assessed for the QC samples at the low, medium, and high concentration levels with the exception of the freeze-thaw tests carried out only for the medium QC level. The stability was evaluated for the samples stored under different conditions: in the autosampler over 24 h at 15 °C, and alternatively, in the refrigerator and the freezer at 4 °C and −18 °C, respectively; exposed to light at room temperature for 1 and 2 h; and four freeze–thaw cycles from −18 °C to room temperature. Three replicates were performed under each condition. Stability was defined as <15% loss of the initial vitamin concentration. The QC samples freshly prepared and measured immediately prior to storage effect testing were used as a reference point.

3.6. Statistical Analysis

The Spearman's rank correlation (rho) was used to estimate the degree of association between the K2MK-7 concentration and the age for a given gender group. The one-way analysis of variance (ANOVA) and *F*-test were used to determine the trending age groups for the given and different gender groups. To determine the significance of each Fisher coefficient (*F*), the *p*-values were used. The values were considered to be significantly different when the result of the compared parameters differed at the $p = 0.05$ significance level. The statistical analysis was performed using Excel (Microsoft Excel 2010).

4. Conclusions

The ample evidence for the relationship between the low vitamin K2MK-7 content in the human body and the occurrence of various dangerous diseases aroused wide interest in this compound both among scientists and average consumers following a healthy diet. As a result, there is an increasing demand for vitamin K determination and a steady increase in the number of samples required to be tested each day. To meet these expectations, in this paper, a simple, efficient, and affordable sample preparation method using the LLME approach, followed by the LC-MS/MS method for the determination of K2MK-7 in the serum samples, was developed. The method was validated according to the FDA guidelines. Then, its feasibility in evaluating K2MK-7 in real samples was tested, proving that it can be adopted as a diagnostic and preventive analytics tool. Under the optimized conditions, using 500 μL of serum and the same amount of n-hexane, the reproducibility and the accuracy were obtained in the ranges of 89–97% and 86–110%, respectively, and the limit of detection value was 0.01 ng/mL. The analysis time for K2MK-7 is about 5 min, with a total analysis time of 10 min necessary to elute more hydrophobic interfering compounds. Based on the results obtained during 2 years, we were able to statistically assess its content and distribution in the Polish population. The obtained Spearman's rank correlation coefficients showed that the K2MK-7 values are significantly related to the age of both females and males. The higher values of the *F*-test obtained for females not only confirmed the relationship between the age and gender with the content of vitamin K, but also indicated that in the female group, there is a greater distribution of K2MK-7 values depending on age. The results will be useful for the subsequent development of K2MK-7 reference intervals in the Polish population.

Supplementary Materials: The following supporting information can be downloaded at: https://www.mdpi.com/article/10.3390/molecules28186523/s1, Figure S1: Calibration curve with marked confidence intervals at the confidence level 0.95.; Figure S2. Exemplary chromatograms obtained during the optimization of sample preparation procedure using different extraction solvents, i.e. CCl_4 (A), CH_2Cl_2 (B), $CHCl_3$ (C), Et_2O (D), and n-hexane (E), respectively. Table S1: Raw data for calibration curve together with statistical analysis; Table S2. The lack-of-fit test; Table S3. The influence of the number of extraction cycles on the surface area of the vitamin K2MK-7 peak extracted using 500 and 250 μL of *n*-hexane, respectively; Table S4. Results of comparative studies.

Author Contributions: Conceptualization, D.W.; Methodology, Ł.P.; Formal analysis, D.S. and I.B. All authors have read and agreed to the published version of the manuscript.

Funding: This research received no external funding.

Institutional Review Board Statement: This study was carried out in accordance with the guidelines of the Polish Council of Medical Research Agencies, which meet the Declaration of Helsinki as revised in 2013. The methodology has been developed for the needs of commercial analyzes performed in a certified diagnostic laboratory. Neither the Polish Center for Accreditation (PN-EN ISO 15189) nor the National Chamber of Laboratory Diagnosticians do not indicate the need to obtain the consent of the bioethics' committee. In general, blood and its fractions, as well as urine, are the most common material for routine determinations in this type of laboratory units. The validation studies, in accordance with the guidelines of the Declaration of Helsinki, involved human volunteers aware of the purpose of the research, from whom informed consent was obtained. Tests on a larger number of blood samples were carried out in accordance with Art. 24 of the Act of 15 September 2022 on laboratory medicine (Journal of Laws, item 2280). "Polish legislation (Act on clinical trials of medicinal products for human use of 13 January 2023) states that ethical approval is not required in this case".

Informed Consent Statement: Not applicable.

Data Availability Statement: Not applicable.

Conflicts of Interest: The authors declare no conflict of interest.

Sample Availability: Samples of the compounds are not available from the authors.

References

1. Sato, T.; Yamada, Y.; Ohtani, Y.; Mitsui, N.; Murasawa, H.; Araki, S. Production of menaquinone (vitamin K2)-7 by Bacillus subtilis. *J. Biosci. Bioeng.* **2001**, *91*, 16–20. [CrossRef] [PubMed]
2. Simes, D.C.; Viegas, C.S.B.; Araújo, N.; Marreiros, C. Vitamin K as a powerful micronutrient in aging and age-related diseases: Pros and cons from clinical studies. *Int. J. Mol. Sci.* **2019**, *20*, 4150. [CrossRef] [PubMed]
3. Walther, B.; Karl, J.P.; Booth, S.L.; Boyaval, P. Menaquinones, bacteria, and the food supply: The relevance of dairy and fermented food products to vitamin K requirements. *Adv. Nutr.* **2013**, *4*, 463–473. [CrossRef] [PubMed]
4. Beulens, J.W.J.; Booth, S.L.; van den Heuvel, E.G.H.M.; Stoecklin, E.; Baka, A.; Vermeer, C. The role of menaquinones (vitamin K2) in human health. *Br. J. Nutr.* **2013**, *110*, 1357–1368. [CrossRef] [PubMed]
5. Paprotny, Ł.; Wianowska, D.; Izdebska, M.; Celejewska, A.; Szewczak, D.; Solski, J. Analysis of serum homocysteine in the laboratory practice—Comparison of the direct chemiluminescence immonoassay and high performance liquid chromatography coupled with fluorescent detection. *Biochem. Med.* **2020**, *30*, 030703. [CrossRef]
6. Fusaro, M.; Gallieni, M.; Porta, C.; Nickolas, T.L.; Khairallah, P. Vitamin K effects in human health: New insights beyond bone and cardiovascular health. *J. Nephrol.* **2020**, *33*, 239–249. [CrossRef]
7. Nimptsch, K.; Rohrmann, S.; Linseisen, J. Dietary intake of vitamin K and risk of prostate cancer in the Heidelberg cohort of the European Prospective Investigation into Cancer and Nutrition (EPIC-Heidelberg). *Am. J. Clin. Nutr.* **2008**, *87*, 985–992. [CrossRef]
8. Dofferhoff, A.S.M.; Piscaer, I.; Schurgers, L.J.; Visser, M.P.J.; van den Ouweland, J.M.W.; de Jong, P.A.; Gosens, R.; Hackeng, T.M.; van Daal, H.; Lux, P.; et al. Reduced Vitamin K Status as a Potentially Modifiable Risk Factor of Severe Coronavirus Disease. *Clin. Infect. Dis.* **2020**, *73*, e4039–e4046. [CrossRef]
9. Wianowska, D.; Bryshten, I. New insights into vitamin K—From its natural sources through biological properties and chemical methods of quantitative determination. *Crit. Rev. Anal. Chem.* 2022; *online ahead of print*. [CrossRef]
10. Knapen, M.H.J.; Vermeer, C.; Braam, L.A.J.L.M.; Theuwissen, E. Pharmacokinetics of menaquinone-7 (vitamin K2) in healthy volunteers. *Clin. Trials* **2014**, *4*, 1000160. [CrossRef]
11. Møller, M.; Fange Gjelstad, I.M.; Baksaas, I.; Grande, T.; Reidun Aukrust, I.; Drevon, C.A. Bioavailability and chemical/functional aspects of synthetic MK-7 vs fermentation-derived MK-7 in randomised controlled trials C.A. *Int. J. Vitam. Nutr. Res.* **2016**, *87*, 297–311. [CrossRef]
12. Klapkova, E.; Cepova, J.; Dunovska, K.; Prusa, R. Determination of vitamins K 1, MK-4, and MK-7 in human serum of postmenopausal women by HPLC with fluorescence detection. *J. Clin. Lab. Anal.* **2018**, *32*, e22381. [CrossRef] [PubMed]
13. Beulens, J.W.J.; Bots, M.L.; Atsma, F.; Bartelink, M.-L.E.L.; Prokop, M.; Geleijnse, J.M.; Witteman, J.C.M.; Grobbee, D.E.; van der Schouw, Y.T. High dietary menaquinone intake is associated with reduced coronary calcification. *Atherosclerosis* **2009**, *203*, 489–493. [CrossRef]
14. Simes, D.C.; Viegas, C.S.B.; Araújo, N.; Marreiros, C. Vitamin K as a diet supplement with impact in human health: Current evidence in age-related diseases. *Nutrients* **2020**, *12*, 138. [CrossRef] [PubMed]
15. Gentili, A.; Cafolla, A.; Gasperi, T.; Bellante, S.; Caretti, F.; Curini, R.; Fernández, V.P. Rapid, high performance method for the determination of vitamin K1, menaquinone-4 and vitamin K1 2,3-epoxide in human serum and plasma using liquid chromatography-hybrid quadrupole linear ion trap mass spectrometry. *J. Chromatogr. A* **2014**, *1338*, 102–110. [CrossRef] [PubMed]
16. Marinova, M.; Lütjohann, D.; Westhofen, P.; Watzka, M.; Breuer, O.; Oldenburg, J. A validated HPLC method for the determination of vitamin K in human serum-first application in a pharmacological study. *Open Clin. Chem. J.* **2011**, *4*, 17–27. [CrossRef]
17. Kamao, M.; Suhara, Y.; Tsugawa, N.; Okano, T. Determination of plasma Vitamin K by high-performance liquid chromatography with fluorescence detection using Vitamin K analogs as internal standards. *J. Chromatogr. B* **2005**, *816*, 41–48. [CrossRef]
18. Karl, J.P.; Fu, X.; Dolnikowski, G.G.; Saltzman, E.; Booth, S.L. Quantification of phylloquinone and menaquinones in feces, serum, and food by high-performance liquid chromatography–mass spectrometry. *J. Chromatogr. B* **2014**, *963*, 128–133. [CrossRef]
19. Riphagen, I.J.; van der Molen, J.C.; van Faassen, M.; Navis, G.; de Borst, M.H.; Muskiet, F.A.J.; de Jong, W.H.A.; Bakker, S.J.L.; Kema, I.P. Measurement of plasma vitamin K1 (phylloquinone) and K2 (menaquinones-4 and -7) using HPLC-tandem mass spectrometry. *Clin. Chem. Lab. Med.* **2016**, *54*, 1201–1210. [CrossRef]
20. Huang, B.; Wang, Z.; Yao, J.; Ke, X.; Xu, J.; Pan, X.-D.; Xu, X.; Lu, M.; Ren, Y. Quantitative analysis of vitamin K 1 in fruits and vegetables by isotope dilution LC-MS/MS. *Anal. Methods* **2016**, *8*, 5707–5711. [CrossRef]
21. Paprotny, Ł.; Celejewska, A.; Frajberg, M.; Wianowska, D. Development and validation of GC–MS/MS method useful in diagnosing intestinal dysbiosis. *J. Chromatogr. B* **2019**, *121822*, 1130–1131. [CrossRef]
22. Gouda, A.S.; Abdel-Megied, A.M.; Rezk, M.R.; Marzouk, H.M. LC-MS/MS-based metabolite quantitation of the antiviral prodrug baloxavir marboxil, a new therapy for acute uncomplicated influenza, in human plasma: Application to a human pharmacokinetic study. *J. Pharm. Biomed. Anal.* **2023**, *223*, 115165. [CrossRef] [PubMed]

23. Rezk, M.R.; Badr, K.A. Development, optimization and validation of a highly sensitive UPLC–ESI-MS/MS method for simultaneous quantification of amlodipine, benazeprile and benazeprilat in human plasma: Application to a bioequivalence study. *J. Pharm. Biomed. Anal.* **2014**, *98*, 1–8. [CrossRef] [PubMed]
24. Rezk, M.R.; Basalious, E.B.; Badr, K.A. Novel determination of sofosbuvir and velpatasvir in human plasma by UPLC–MS/MS method: Application to a bioequivalence study. *Biomed. Chromatogr.* **2018**, *32*, e4347. [CrossRef] [PubMed]
25. Viñas, P.; Bravo-Bravo, M.; López-García, I.; Hernández-Córdoba, M. Dispersive liquid–liquid microextraction for the determination of vitamins D and K in foods by liquid chromatography with diode-array and atmospheric pressure chemical ionization-mass spectrometry detection. *Talanta* **2013**, *115*, 806–813. [CrossRef] [PubMed]
26. Rishipal, S.; Alka, P.; Mojeer, H.; Bibhu Prasad, P. Development of a Rapid HPLC-UV Method for Analysis of Menaquinone-7 in Soy Nutraceutical. *Pharm. Anal. Acta* **2016**, *7*, 7–12. [CrossRef]
27. Wianowska, D.; Gil, M.; Olszowy, M. Miniaturized methods of sample preparation. In *Handbook on Miniaturization in Analytical Chemistry: Application of Nanotechnology*; Elsevier: Amsterdam, The Netherlands, 2020; pp. 99–125. [CrossRef]
28. Agilent Technical Overview: Making Your LC Method Compatible with Mass Spectrometry, Edgar Naegale, Waldbronn, Germany. Available online: https://community.agilent.com/cfs-file/__key/docpreview-s/00-00-00-85-46/5990_2D002D00_7413EN.pdf (accessed on 12 August 2022).
29. Colizza, K.; Mahoney, K.E.; Yevdokimov, A.V.; Smith, J.L.; Oxley, J.C. Acetonitrile ion suppression in atmospheric pressure ionization mass spectrometry. *J. Am. Soc. Mass Spectrom.* **2016**, *27*, 1796–1804. [CrossRef]
30. Szterk, A.; Bus, K.; Zmysłowski, A.; Ofiara, K. Analysis of Menaquinone-7 Content and Impurities in Oil and Non-Oil Dietary Supplements. *Molecules* **2018**, *23*, 1056. [CrossRef]
31. U.S. Department of Health and Human Services, Food and Drug Administration. *Guidance for Industry: Bioanalytical Method Validation*; FDA: Silver Spring, MD, USA, 2001; pp. 4–10. Available online: http://www.fda.gov/downloads/Drugs/Guidances/ucm070107.pdf (accessed on 12 August 2022).
32. Vermeer, C.; Raes, J.; van 't Hoofd, C.; Knapen, M.H.J.; Xanthoulea, S. Menaquinone content of cheese. *Nutrients* **2018**, *10*, 446. [CrossRef]
33. Matuszewski, B.K.; Constanzer, M.L.; Chavez-Eng, C.M. Strategies for the Assessment of Matrix Effect in Quantitative Bioanalytical Methods Based on HPLC−MS/MS. *Anal. Chem.* **2003**, *75*, 3019–3030. [CrossRef]
34. U.N. Department of Economic and Social Affairs. The Average Age in Global Comparison. Available online: www.worlddata.info/average-age.php (accessed on 12 August 2022).
35. Hu, K.; Li, Y.; Ding, R.; Zhai, Y.; Chem, L.; Qian, W.; Yang, J. A simple, sensitive, and high-throughput LC-APCI-MS/MS method for simultaneous determination of vitamin K1, Vitamin K1 2,3-epoxide in human plasma and its application to a clinical pharmacodynamics study of warfarin. *J. Pharm. Biomed. Anal.* **2018**, *159*, 82–91. [CrossRef]
36. Dunovska, K.; Klapkova, E.; Sopko, B.; Cepova, J.; Prusa, R. LC–MS/MS quantitative analysis of phylloquinone, menaquinone-4 and menaquinone-7 in the human serum of a healthy population. *PeerJ* **2019**, *7*, e7695. [CrossRef] [PubMed]

molecules

MDPI

Article

Epoxidation of Methyl Esters as Valuable Biomolecules: Monitoring of Reaction

Martin Hájek [1], Tomáš Hájek [2,*], David Kocián [1], Karel Frolich [1] and András Peller [3]

[1] Department of Physical Chemistry, Faculty of Chemical Technology, University of Pardubice, Studentská 95, 532 10 Pardubice, Czech Republic; martin.hajek@upce.cz (M.H.); david.kocian@student.upce.cz (D.K.); karel.frolich@upce.cz (K.F.)
[2] Department of Analytical Chemistry, Faculty of Chemical Technology, University of Pardubice, Studentská 95, 532 10 Pardubice, Czech Republic
[3] Faculty of Chemical and Food Technology, Slovak University of Technology in Bratislava, Radlinského 9, 81237 Bratislava, Slovakia; andras.peller@stuba.sk
* Correspondence: tomas.hajek@upce.cz

Abstract: The paper is focused on the epoxidation of methyl esters prepared from oil crops with various profiles of higher fatty acids, especially unsaturated, which are mainly contained in the non-edible linseed and *Camelina sativa* oil (second generation). The novelty consists in the separation and identification of all products with oxirane ring formed through a reaction and in the determination of time course. Through the epoxidation, many intermediates and final products were formed, i.e., epoxides with different number and/or different position of oxirane rings in carbon chain. For the determination, three main methods (infrared spectroscopy, high-pressure liquid chromatography and gas chromatography with mass spectrometry) were applied. Only gas chromatography enables the separation of individual epoxides, which were identified on the base of the mass spectra, molecule ion and time course of products. The determination of intermediates enables: (i) control of the epoxidation process, (ii) determination of the mixture of epoxides in detail and so the calculation of selectivity of each product. Therefore, the epoxidation will be more environmentally friendly especially for advanced applications of non-edible oil crops containing high amounts of unsaturated fatty acids.

Keywords: vegetable oils; epoxidation; esters; gas chromatography; infrared spectroscopy; liquid chromatography; gas chromatography

1. Introduction

The current global issue is the searching for renewable sources for production of various materials, chemicals, or energy, which are currently produced from crude oil. Triacylglycerides, contained in the vegetable oils, animal fats or waste frying oils [1], are one of the possible renewable sources and can be transformed to esters by transesterification [2]. An important product is glycerol, which has many applications in chemistry, food and pharmaceutical industries [3] and is produced especially from oils/fats. The esters are mainly used as a fuel, but can be also transformed to epoxides, which have many applications. The epoxidation is a reaction of double bonds between two carbon atoms with hydrogen peroxide which leads to the formation of the epoxide group [4]. The epoxides can be used as: (i) bio-lubricants in means of transport (additives to oils) [5] or (ii) raw materials for bio-polymers (as monomers), higher alcohols, olefins, glycols, polyesters and polycarbonates (reaction with CO_2, which is consumed) [6,7]. These chemicals are currently produced from crude oil and it is appropriate to replace them by renewable sources.

The different oils/fats have different profiles of higher fatty acids (FA) in triacylglyceride, which influence the physical and chemical properties of oil and products such as freezing point, oxidation stability, and kinematic viscosity [8]. The most frequent fatty acids in common oils/fats are the following: palmitic (16:0), stearic (18:0), oleic (18:1), linoleic (18:2), and linolenic (18:3) [9]. Note: the first number is the number of carbon atoms and the second is the number of double bonds. The FA profile of FA-formed esters is the same as in oils/fats, i.e., does not change during transesterification [10]. For epoxidation, the oils with higher content of unsaturated FA, i.e., higher iodine value (IV) are more suitable, because: (i) higher amounts of epoxides are formed, and (ii) these raw materials cannot be used as fuels (maximum IV is 120 g I_2/100 g according to EN 14214 because of low oxidation stability). For this reason, the oil from non-edible oil crops, i.e., "second generation" is appropriate because it usually has a high content of unsaturated FA [11] and, moreover, can be used for the production of glycerol as a by-product by transesterification. This is differed from the epoxidation of oil, where glycerol is not formed.

The monitoring of the epoxidation reaction (especially time dependency) is important because it allows to control the course of epoxidation in detail. The unsaturated esters of higher FA (usually containing 1–3 double bonds depending on the type of acid) are consecutively epoxidated, therefore many intermediates are formed. Moreover, the esters can form two geometric isomers (cis or trans) [12], which increase the number of products. Produced epoxides can also react with water, hydrogen peroxide or organic acid to form alcohols [13]. Several monitoring methods are usually used, such as determination of iodine value, the epoxide equivalent weight, epoxy index [14], kinematic viscosity, infrared spectroscopy [15], and flash and combustion points [16]. However, the methods determine only the change of double bonds or change of content of oxirane groups without detailed identification of the type of epoxides. The ^{1}H NMR method can separate the substances according to the number of oxirane rings, but it is very often used for triglycerides (not for esters) [17,18]. On the other hand, the chromatographic methods allow to separate esters according to FA, including cis or trans isomers [19].

The major advantage of gas chromatography (GC) is the ability to determine all compounds in the reaction mixture individually in one analysis. The GC is easy, highly sensitive, automated, and provides analysis of data which gives comparatively higher precision, accuracy, and reproducible results. The GC is usually applied for determination of alkyl esters with a different number of double bonds (different type of acid). However, the method is not commonly used for separation of esters with oxirane ring and esters without it [20]. Methyl esters and their epoxides can be identified by comparison of their retention times with standards or can be identified by using mass spectrometry (MS) on the base of their mass spectrum (intensity as a function of the mass-to-charge ratio, m/z). The fatty acids or methyl esters of fatty acids with oxirane rings are generally very poorly studied and there is a lack of information regarding their fragmentation by electron ionization.

The course of epoxidation of methyl esters was determined by two chromatographic methods (GC with mass spectrometry detector and high-pressure liquid chromatography with refractometric detector), which were compared mutually and with another method (infrared spectroscopy). The novelty consists in the identification of many formed intermediates (various degrees of epoxidation) including their reaction time dependency (no study focused on their determination has been carried out yet). Moreover, various esters with different compositions of higher fatty acids were used, i.e., with different amounts of double bonds. Detailed knowledge of the reaction course will allow to control the epoxidation process better, and thus reduce the reaction cost and raw material consumption, especially for non-edible oils with high contents of unsaturated FA.

2. Results and Discussion

The esters of vegetable oils with different profiles of fatty acids were used, such as rapeseed oil, sunflower oil and linseed oil. Their composition was determined using GC-MS (Table 1 includes iodine value and water content). These oils contain saturated (stearic and palmitic) and especially unsaturated (oleic, linoleic and linolenic) fatty acids, which were epoxidized. Note: only the most frequently occurring higher fatty acids, which represent approximately 98% of content, are stated in Table 1.

Table 1. The composition and profiles of oils. IV—iodine value.

Type of Oil (Abbreviation)	IV (g I_2/100 g)	Water Content (ppm)	Profile of Higher Fatty Acids [1] (wt.%)							
			16:0	18:0	18:1	18:2	18:3	20:1	20:2	22:1
Rapeseed (RO)	106.9	260	6.4	2.7	55.9	25.5	6.7	1.2	- *	0.1
Sunflower (SO)	111.4	440	8.0	4.8	41.1	44.6	-*	0.1	- *	- *
Linseed (LO)	181.3	350	7.6	5.3	23.6	17.8	44.3	0.1	- *	- *
Camelina sativa (CS)	151.3	430	7.1	2.9	16.9	19.8	27.6	16.2	1.6	3.3

[1] number of carbon atoms: number of double bonds. * not detected.

Throughout epoxidation, many chemical substances were formed such as: (i) raw materials (esters), (ii) reaction intermediates and (iii) final products of epoxidation (Table 2). The intermediates are polyunsaturated esters with various degrees of epoxidation, i.e., various numbers of oxirane rings (one or two) or esters with oxirane rings placed on different positions in the methyl esters chain. Moreover, the esters and epoxides with double bonds can be present in the form of cis and trans isomers. The standards of these substances (intermediates and epoxides) are not commercially available.

2.1. ^{1}H NMR

The method ^{1}H NMR is applied for the analysis of epoxides, except in most cases for (i) the epoxidation of oils (not esters) and (ii) start and end of the reaction (not reaction course). Moreover, authors usually determine only the decrease in signal of double bonds between carbon atoms (C=C) or increase of the oxirane group [21–23]. Similar, epoxides with one, two, or three oxirane rings were identified [24]. Moreover, some authors used just one pure substance (standard) as a raw material such as oleic acid, trilinolein, trilinolenin, etc., and so only few (often one) products had been formed [24,25]. Mushtaq at al. used ^{1}H NMR for epoxidation of fatty acids and methyl ester of Jatropha oil, but also only determined the signal of C=C bonds and oxirane without detailed identification [26]. Xia W. et al. and Goicoechea E. et al. were successful in the identification of each individual signal of the ^{1}H-NMR spectrum for the epoxidation of sunflower, canola and fish oil. The authors used the signal of glycerol backbone structure as an internal reference for signal correction; however, it cannot be used in the mixture of epoxidized esters because it does not contain glycerol.

The spectrum of methyl esters and epoxidized methyl esters of *Camelia sativa* oil was determined (Figure S1 in Supplementary Materials). Neither the level of epoxidation, nor quantification of each product can be made, because the signals of mono/di/tri-epoxides cannot be assigned to individual molecules, which is the same as in paper [24]. Therefore, ^{1}H NMR is not appropriate for the identification of each individually formed epoxide (Table 2), only for confirmation of the oxirane group.

2.2. The Simulated Distillation

The simulated distillation method allows the separation of high boiling-temperature chemical substances based on their boiling point [27]. The result of this analysis is the dependence of the weight fraction on the boiling point for the methyl ester (ME) and epoxide methyl ester (E_ME) (Figure 1).

For ME, only two peaks were determined with the maximum temperature: (i) 325 °C, which corresponds to ester containing 16 carbons (ester of saturated palmitic acid), and (ii) 360 °C for esters containing 18 carbons (oleate, stearate, linoleic and linolenic), which have higher boiling points due to their higher number of carbon atoms [19]. The boiling point of esters corresponded with tabulated data [28]. For E_ME, more peaks were observed: (i) the first peak at lower temperatures (325 °C) was unreacted palmitic esters, (ii) the peak at 360 °C corresponded to unreacted esters containing 18 carbons, and (iii) peaks at 375 and 395 °C were attributed to epoxides, because the oxygen increased the boiling point of esters. The splitting of the peaks at 395 °C is due to a different number of oxirane rings in methyl esters of polyunsaturated fatty acids, which corresponds to different types of epoxides (Table 2) with slightly different boiling points. This method is not sensitive enough to differentiate each type of ester or epoxide and is thus not suitable for the analysis of reaction mixtures in detail.

Table 2. The products of epoxidation from the most common unsaturated esters.

Type of Fatty Acid Methyl Ester		Type of Epoxy Fatty Acid Methyl Ester	M (g/mol)
cis-9-octadecenoic (oleic) acid ME 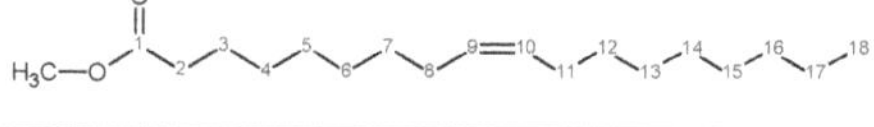	*Monoepoxy (C18:1-Ep)*	*cis*-9,10-epoxy octadecanoate ME 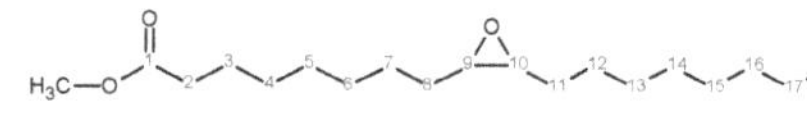	312.5
cis,cis-9,12-octadecenoic (linoleic) acid ME 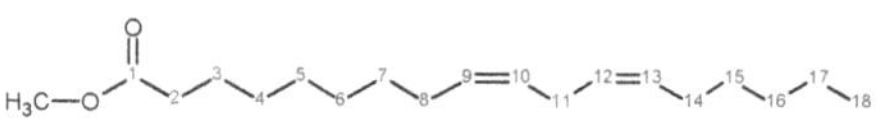	*Monoepoxy (C18:2 1-Ep)*	*cis*-9,10-epoxy octadec-12-enoate ME 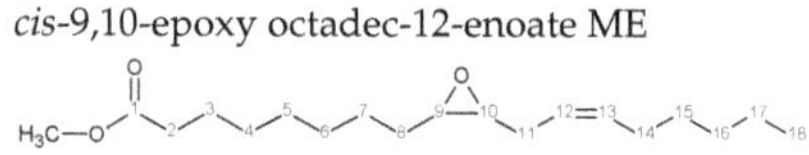 *cis-12,13*-epoxy octadec-9-enoate *ME* 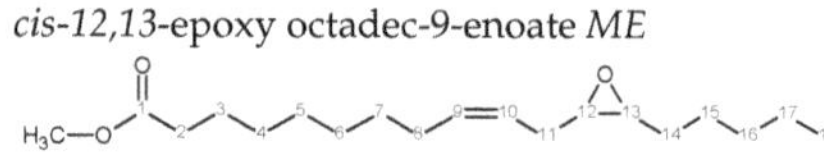	310.5
	Diepoxy (C18:2 2- Ep)	*cis,cis*-9,10;12-13-diepoxy octadecanoate *ME* 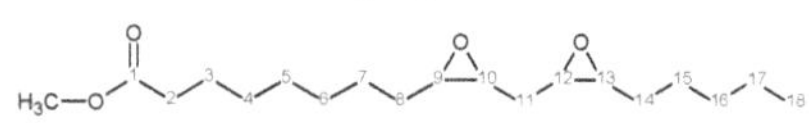	326.5
cis,cis,cis-9,12,15-octadecenoic (linolenic) acid ME 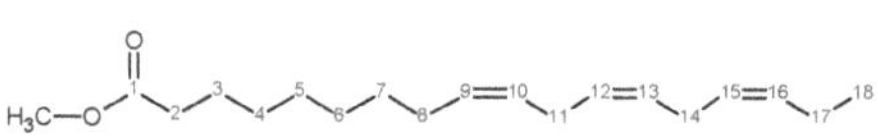	*Monoepoxy (C18:3 1-Ep)*	*cis*-9,10-epoxy octadec-12,15-dienoate ME 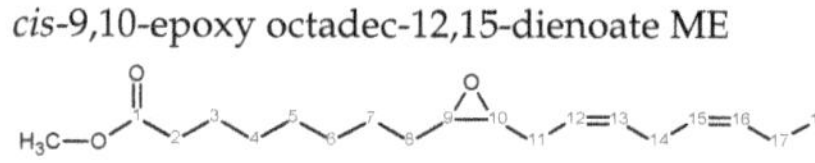 *cis*-12,13-epoxy octadec-9,15-dienoate ME 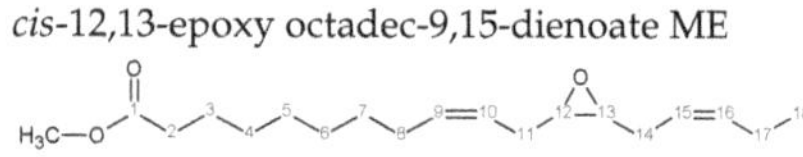 *cis*-15,16-epoxy octadec-9,12-dienoate ME 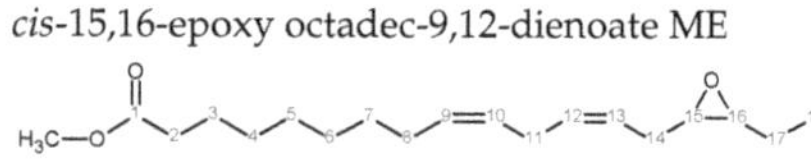	308.5
	Diepoxy (C18:3 2-EPO)	*cis,cis*-9,10;12,13-diepoxy octadec-12-enoate ME 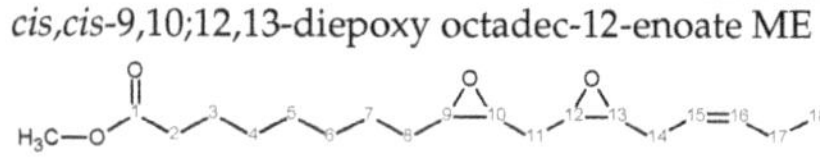 *cis,cis*-12,13;15,16-diepoxy octadec-9-enoate ME 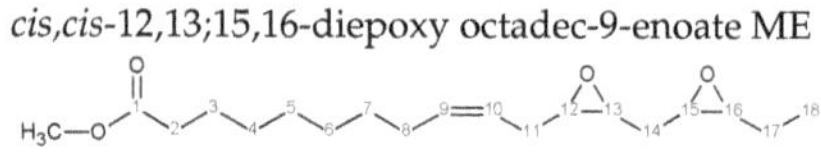 *cis,cis*-9,10;15,16-diepoxy octadec-12-enoate ME 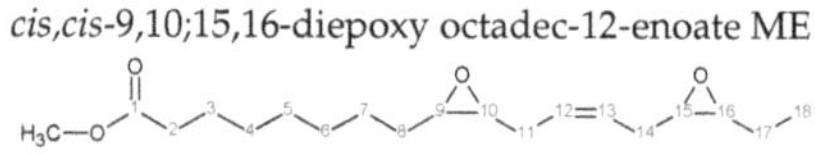	324.5
	Triepoxy (C18:3 3-Ep)	*cis,cis,cis*-9,10;12,13;15,16-triepoxy octadecanoic acid ME 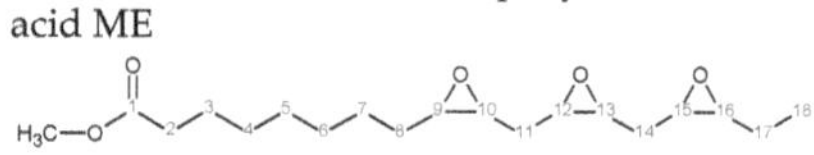	340.5

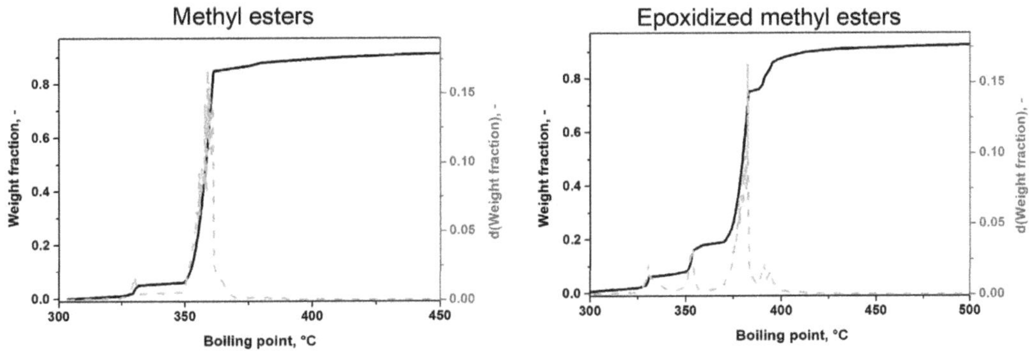

Figure 1. The plot of the simulated distillation for methyl esters and epoxidized methyl esters.

2.3. The Infrared Spectroscopy

The infrared spectra of the reaction mixture during the reaction time (24 h) were determined for esters of sunflower and linseed oil (Figure 2). Furthermore, the spectra for esters of rapeseed oil and *Camelina sativa* oil were also determined; however, they are not presented because they were almost the same as for rapeseed oil and linseed oil, respectively.

The absorption band with a maximum at 1743–1740 cm^{-1} was attributed to valence vibrations of the methyl ester carbonyl group (Figure 2A) [29,30]. A slight shift towards smaller wavelengths for the ME of sunflower and linseed oil were observed during the first two hours of epoxidation, which can be explained by the formation of an oxirane ring in the molecule influencing the strength of the bond. For ME_LO, the absorption band with the maximum at 1640 cm^{-1} may belong to deformation vibrations H-O-H from the water that may be present in the samples. The bands with a maximum at 2926 cm^{-1} were attributed to valence vibrations in the groups of alkanes and alkenes (Figure 2B). The band with the maximum at 3010 cm^{-1} was attributed to valence vibrations of double bonds between carbons and disappeared within the first three hours of epoxidation. Simultaneously, the intensity of three bands at 2954, 2926 and 2855 cm^{-1} increased, corresponding to a single bond between carbons. Therefore, the double bonds between two carbons were transformed to the single bonds in accordance with the epoxidation. This was confirmed by the decrease in the iodine value from 111.7 to 1.0 g $I_2/100$ g (for ME_SO) and from 181.3 to 5.5 g $I_2/100$ g (for ME_LO).

For ME_LO, the signal belonging to the valence vibrations of OH groups (maximum at 3417 cm^{-1}) increased with increasing reaction time (Figure 2A). This indicates the subsequent hydrolysis of the epoxides to alcohols [31]. The hydrolysis corresponds with (i) almost zero EI, i.e., almost zero oxirane rings and (ii) a high kinematic viscosity (450 mm^2/s) after 24 h of reaction. Moreover, the alcohol formation has already been described for higher reaction temperatures and longer reaction times [32]. The hydrolysis does not occur during the epoxidation of ME_SO (a band with maximum 3417 cm^{-1} was not present), because only a small amount of methyl ester of linolenic acid is present in SO. Therefore, high content of ME of linolenic acid caused the formation of alcohols because it is the most reactive ester due to the presence of three double bonds [22].

In the area of fingerprint (Figure 2C), the differences between the following absorption bands were determined: (i) the deformation vibrations of alkanes with a single bond (1460 and 1370 cm^{-1}) and (ii) the valence vibrations of the oxirane ring (824–845 cm^{-1}). Both bands increased in the first two hours of the reaction because the oxirane ring with single bonds was formed. Other bands with the maximum at 1171, 1198 and 1248 cm^{-1} were attributed to the valence vibrations C-O in the ester functional group and did not change during the epoxidation. For ME_LO, the band of valence vibrations C-O in alcohol (maximum at 1072 cm^{-1}) increased with increasing reaction time (similar to O-H band with a maximum at 3417 cm^{-1}). This verified the opening of the oxirane ring and alcohol formation already after three reaction hours for ME_LO.

The infrared spectroscopy determined the change of chemical bonds during the reaction time but was not able to identify individual ME with different oxirane rings.

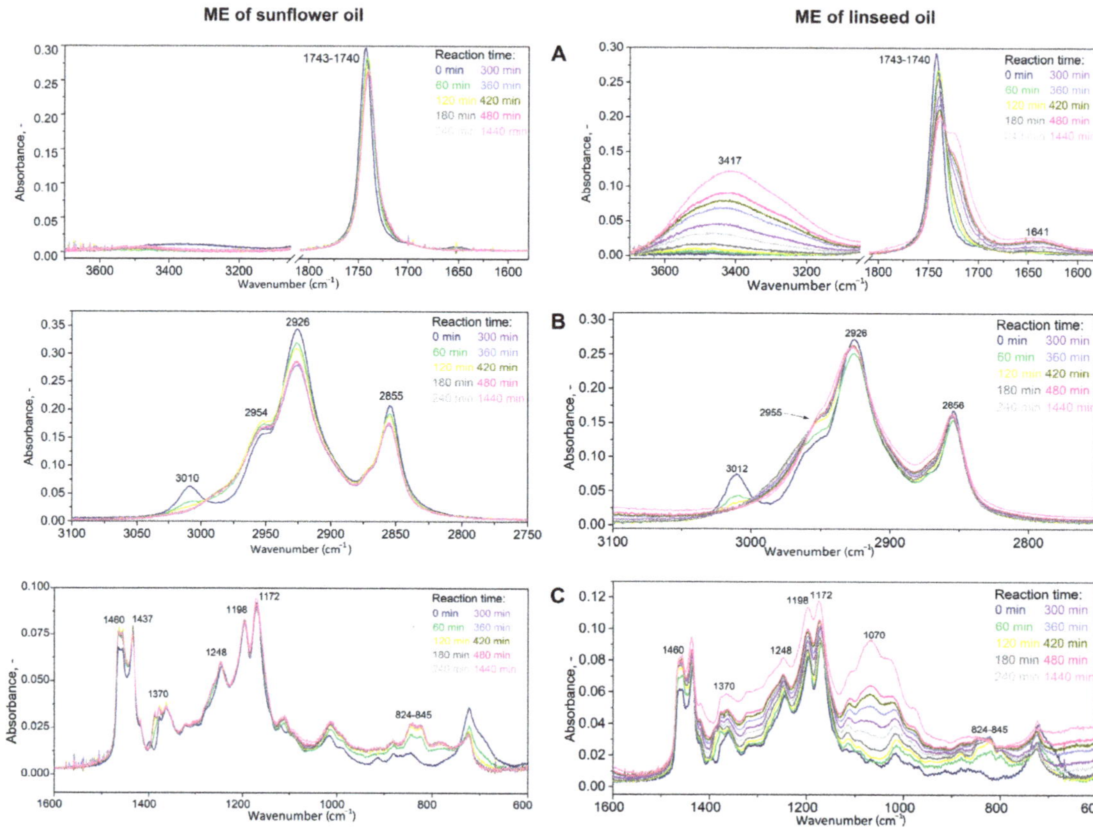

Figure 2. The infrared spectra of the reaction mixture during epoxidation for ME of sunflower and linseed oil. (**A**)—wavenumber range of 3700–1600 cm^{-1}, (**B**)—wavenumber range of 3100–2750 cm^{-1}, (**C**)—wavenumber range of 1600–600 cm^{-1}.

2.4. High Performance Liquid Chromatography with Refractometric Detection

The HPLC-RI method is based on the analysis of esters in the glycerol phase after transesterification [33]. However, the method parameters had to be significantly modified and optimized, such as: (i) the type of stationary phase in the column, (ii) the type of mobile phase (acetonitrile and methanol including their various volume ratios), (iii) the mobile phase flow rate (0.2–0.7 mL/min) and (iv) the column temperature (27–45 °C).

Several columns with different reverse-phase stationary phases and with different dimensions were tested for the separation of esters with and without the oxirane ring. The best results were obtained with the YMC Carotenoid C30 column and the C18 column connected in series. The use of only one of these columns alone resulted in the co-elution of part of the compounds (the ME of linolenic acid had the same retention time as its ME with oxirane ring). The pure methanol at a flow rate of 0.7 mL/min was better as a mobile phase than pure acetonitrile or various ratios of these solvents even though the column back pressure was higher. Complete separation was performed in 25 min. The higher temperature (35–45 °C) had a negative effect on the separation quality and peak shape, so the lowest temperature (27 °C) was chosen. The dependency of intensity of peaks on reaction time for ME of sunflower and linseed oils was determined at optimized conditions (Figure 3).

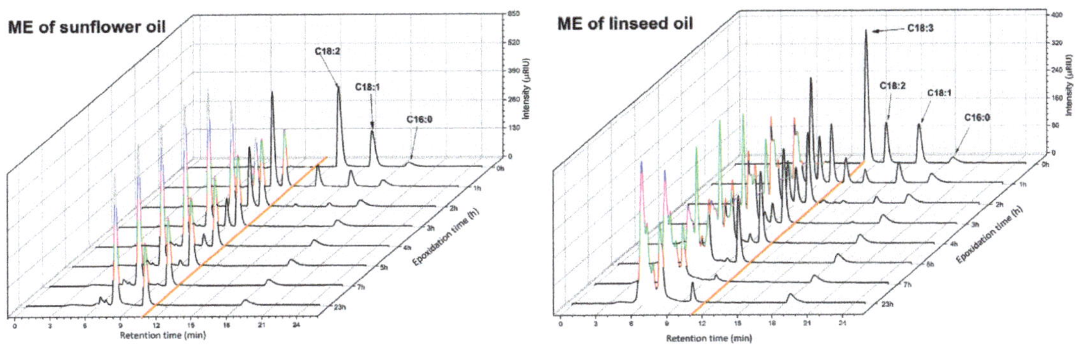

Figure 3. The HPLC chromatogram of reaction mixture during the epoxidation for ME of sunflower and linseed oil (orange line separates ME with and without oxirane rings).

The methyl esters with oxirane rings were eluted in the retention time of 5.0–11.4 min. The methyl esters without oxirane ring were eluted later (retention time 11.5–25.0 min) and were clearly identified by standards. Their retention time increased with increasing Equivalent Carbon Number (ECN) of the ester molecule in the following order: ME of linolenic acid, linoleic acid, oleic acid and palmitic acid. Note: ECN is calculated as the number of carbons in the chain minus two times the number of double bonds [34]. However, the ME with oxirane rings were not clearly separated, especially for linseed oil, where more intermediates were formed.

The HPLC method was sufficient for the separation of ME with and without oxirane rings and was able to determine the conversion of ME to their epoxides. On the other hand, this method was not able to separate the ME with a different number or position of oxirane rings, regardless of the number of carbons. Anuar at al. published the epoxidation of triglyceride and identified the total number of double bonds in molecule by Liquid Chromatography/Mass Spectrometry, but the method did not readily distinguish between positions of oxirane ring and isomers [35].

2.5. Gas Chromatography with Mass Spectrometry

The dependency of the reaction mixture composition on time was determined by GC-MS (typical chromatogram is in Figure 4). The C17:0 ME as the internal standard was added to all samples to help improve the accuracy of the composition. The peaks of ME without an oxirane ring were identified on the base of their mass spectrum, which is a plot of the ion-relative abundance versus mass-to-charge ratio (*m/z*). The mass spectra at the peak apex were compared with the NIST Mass Spectra Library.

All methyl esters without an oxirane ring were clearly identified and their retention times are in Table S1 in Supplementary Materials. The content of saturated ME was approximately constant in the reaction time, while the decrease of unsaturated ME during the reaction time was observed, which was expected (Figure 5(A1–C1)). The methyl esters with oxirane ring(s) are assigned according to the number of atoms in the carbon chain and the number of double bonds before epoxidation; then, the number of the epoxide group and the last roman numbering indicate the variant of epoxides (different position of oxirane group or cis/trans), i.e., C18:3 2-Ep I is epoxide from ester of linolenic acid (C18:3) with two oxirane group (2-Ep) and some cis/trans combination (I).

The peak at 32.4 min was identified by the NIST library as the methyl ester of oleic acid with one oxirane ring (C18:1 1-Ep), i.e., fully epoxidized. The peaks with higher retention times were other ME with oxirane rings formed during epoxidation, i.e., various degree of epoxidation (Table 2). Unfortunately, the NIST library does not contain mass spectra of other ME with oxirane rings (epoxides) and standards are not available yet. Their retention times were higher than for the methyl ester of oleic acid with one oxirane ring, which corresponded with: (i) a higher boiling point of methyl esters with a higher number of oxygens in the molecule, and (ii) results of SimDis (Figure 1). Therefore, the oils with different profiles of fatty acids were chosen to identify other peaks; namely: the rapeseed oil with a high content of C18:1, sunflower oil with a high content of C18:2, linseed oil with a higher content of C18:3 and *Camelina sativa* with high content of C20:1 (Table 1). The dependency of each peak (the most intensive ion) of the reaction mixture on time was determined (Figure 5).

For the methyl ester of rapeseed oil, the first most intensive peak from the epoxides group was attributed to oleic methyl ester with one oxirane group. This also confirmed the dependency of intensity on reaction time: the intensity increased to 24 h of reaction and then slightly decreased, which was caused by side reaction [13]. The EI was quite high, at 3.2 mol/kg after 24 h of reaction.

For the methyl ester of sunflower oil (Figure 5B2), two peaks (retention time 33.3 and 34.5 min) with similar intensity and time course were identified on the base of mass spectra of 12,13-epoxy-octadec-9-enoate and 9,10-epoxy-octadec-12-enoate [36]. These two peaks are ME of linoleic acids with one double bond transformed to the oxirane group and one double bond remaining (signed as C18:2 1-Ep I and II). Moreover, the maximum intensity of signal was detected after only 60 min of reaction time and then still decreased, which indicated reaction intermediates. The peaks with retention time 49.2 and 52.2 min were identified as fully epoxidized ME of linoleic acid (contains two oxirane groups, signed as C18:2 2-Ep I and II). The reason is the formation of (i) the molecular ions (*m/z* 326), which corresponded with molar mass of fully epoxidized ME of linoleic acid (Table 2), and (ii) the increasing of peak intensity to 8 h of the reaction time and then remaining almost stable, i.e., the formation of stable final products. These two epoxides are differed by geometric conformation (cis and trans). The C18:2 2-Ep with lower retention time (49.2 min) is probably a trans isomer, because trans isomers have a lower boiling point than cis isomers [37]. Moreover, the formation of epoxides was confirmed by a very low iodine value (1.0 g I_2/100 g) and a high epoxide index (3.6 mol/kg) in the final products.

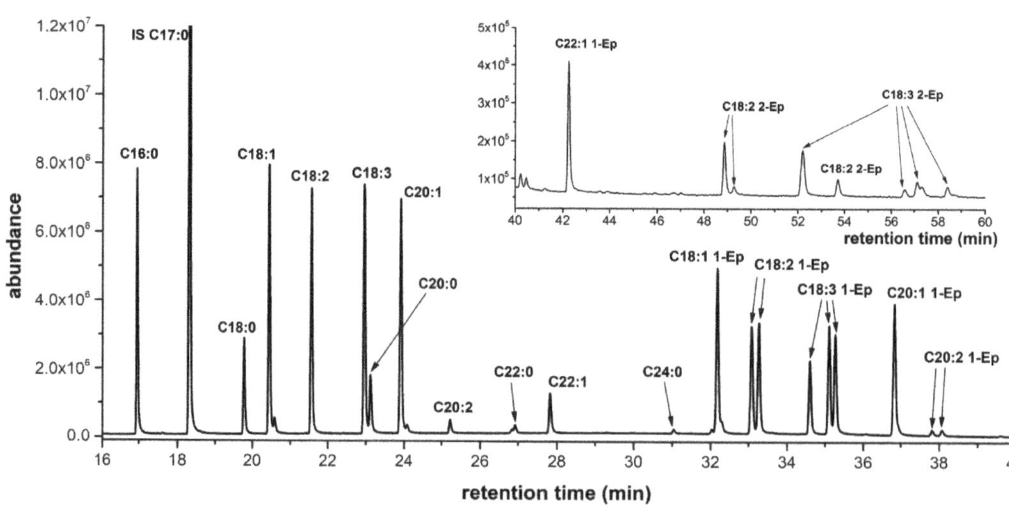

Figure 4. Typical chromatogram of reaction mixture. The chromatogram of the reaction mixture of *Camelina sativa* after 60 min of reaction.

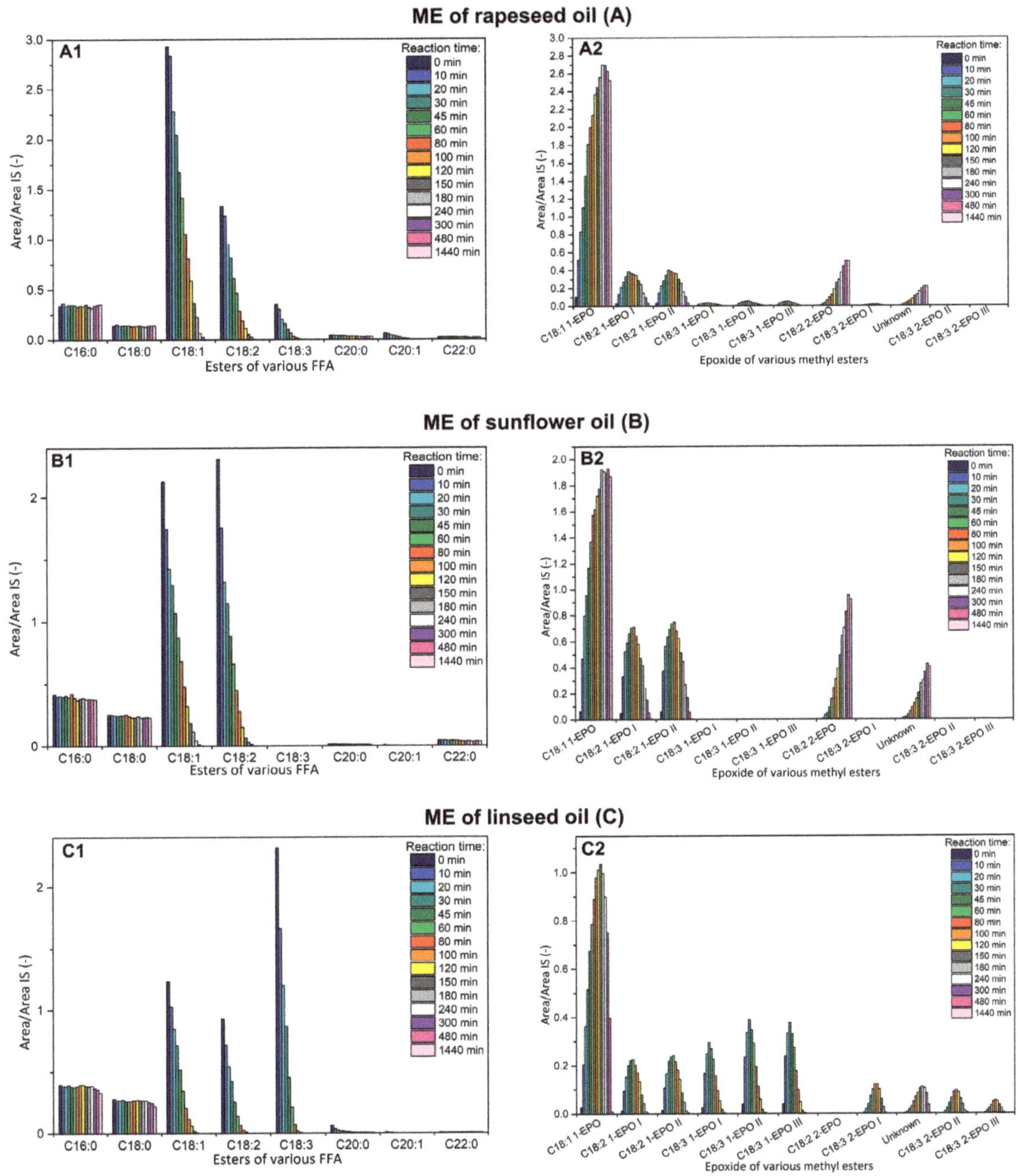

Figure 5. The dependency of response ratio (A/A_{IS}) of methyl esters of FA (**A1–C1**) and epoxides of methyl esters of FA (**A2–C2**) on the reaction time for epoxidation of ME of rapeseed (**A**), sunflower (**B**) and linseed oil (**C**).

For the methyl ester of linseed oil, the most intensive peaks were linolenic methyl esters and therefore many intermediates can be formed during the reaction. The three peaks with retention times 34.8, 35.3 and 35.5 min were attributed to the ME of linolenic acid with one oxirane group and two double bonds (three possibilities, Table 2), signed as C18:3 1-Ep I, II and II. The peaks were identified on the base of total ion current and molecular ions (*m/z* 308), which corresponded with the molar mass of epoxide. Generally, the retention time (order of peaks) increases (i) firstly with increasing number of oxirane group, (ii) secondly with increasing number of carbons, and (iii) thirdly with the number of double bonds. Therefore, their retention time was higher than C18:1 1-Ep and C18:2 1-Ep. The reason is that the presence of oxygen as well as number of double bonds increases the boiling point. Moreover, the dependency of all peaks C18:3 1-Ep on reaction time also corresponds with the formation of intermediates: the highest intensity was after 45 min and then decreased, because the other double bond was transformed to oxirane. The peaks with close retention time (35.3 and 35.5 min) probably correspond to C18:3 1-Ep with oxirane at the position 9,10 and 15,16 because of similar structure (one oxirane ring and two double bonds adjacent, Table 2). While the peak with 35.5 min corresponds with C18:3 1-Ep with oxirane at 12,13 positions.

The four peaks with retention times 52.2, 56.8, 57.1 and 58.4 min and the same time course were attributed to the ester of linolenic acid with two oxirane groups and one double bond, signed as C18:3 2-Ep I, II, III and IV (determination of molecular ion 324). These peaks possess various combinations of oxirane group positions and geometrical isomers. The four peaks of epoxides were also found after the epoxidation of trilinolenin (without identification in detail) [24]. Moreover, the maximum intensity was at 100 min of reaction, i.e., about 55 min later than for C18:3 1-Ep, which also confirms the formation of epoxides with two oxirane rings by consecutive reaction. The intensity of all peaks C18:3 2-Ep decreased after 100 min, which was caused by epoxide ring opening (alcohols were formed, which was confirmed by infrared spectroscopy and almost zero EI); this is the reason why the methyl ester of linolenic acid with three oxirane groups was not determined.

The esters formed from non-edible *Camelina sativa* oil contain more types of other higher fatty acids (C20:1, C20:2 and C22:1), which caused more types of formed epoxides (Figure 6) and so the presence of new peaks. A quite intensive peak with retention time 36.8 was assigned to C20:1 1-Ep. Two small similar peaks with closer retention times 37.8 and 38.1 min were reaction intermediates of C20:2 (C20:2 1-Ep I and II), because they reach maximum concentration after 120 min of reaction, i.e., one oxirane ring and one double bond. The course of these three peaks is similar to C18:1 1-Ep and C18:2 1-Ep but with higher retention time (retention time increases with increasing number of carbons). Another peak with retention time 42.2 min was probably C22:1 1-Ep, because of the same time course as C18 1-Ep and C20 1-Ep. In the end of the epoxidation, the epoxy index was quite small (1.66 mol/kg), which indicated the opening ring reaction and so decrease of the content of epoxides within the last several hours of reaction.

However, GC is not an appropriate method for the determination of ring opening reaction products, especially alcohols, because of their high boiling point (between 450–650 °C depending on the number of alcohols groups [38]). The appropriate method is HPLC with reverse phase [39] or infrared spectroscopy (Section 2.3).

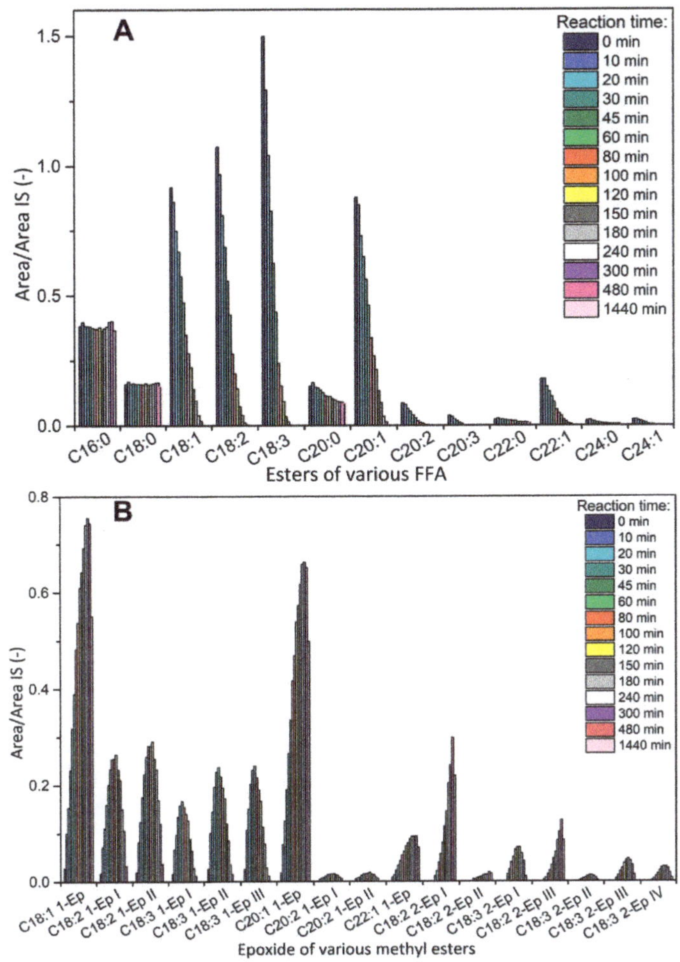

Figure 6. The dependency of the response ratio (A/A_{IS}) of ME from *Camelina sativa* oil on the reaction time for esters (**A**) without oxirane and (**B**) with oxirane ring.

3. Materials and Methods

3.1. Epoxidation of Esters of Higher Fatty Acids

The methyl esters were prepared by transesterification from various types of oil and methanol by standard methods [40]. The epoxidation of formed esters was carried out by hydrogen peroxide (30%, technical, Lach-Ner, Neratovice, Czechia) formic acid and sulfuric acid (pure, Lach-Ner) as a catalyst. The amount of 326 mL of methyl esters was mixed with 31.4 mL of formic acid (molar ratio of formic acid to double bonds was 0.7:1) in a batch reactor and 0.358 mL of 1 wt.% sulfuric acid was added. The mixture was cooled to 8–10 °C and 266 mL of hydrogen peroxide (the molar ratio of hydrogen peroxide to double bonds was 2:1) was added gradually over 30 min at stirring speed 300 rpm. After the addition of hydrogen peroxide, the temperature was increased to 60 °C and the reaction proceeded for 24 h. The reaction conditions were chosen based on the review [2]. The amount of 5 mL of the reaction mixture was sampled at time intervals and washed by 8 mL of solution of potassium carbonate (4.5 wt.%) to neutralize the acids, and then by 8 mL of water. The water excess was removed by decantation and centrifugation (4000 rpm for 20 min) and individual samples were then analyzed.

The rest of the reaction mixture (after 24 h) was washed several times with a potassium carbonate solution until the pH of the epoxide phase was neutral, and the aqueous phase was removed by decantation in a separatory funnel. Subsequently, the residual water was removed by distillation with the addition of methanol [41].

3.2. Analytical Methods

The content of methyl esters with or without the oxirane ring were determined by three main methods during the reaction.

Infrared spectroscopy: The infrared spectroscopy (FTIR) with ATR module—diamond (Nicolet™ iS50 FTIR Spectrometer, Thermo Fisher Scientific, Waltham, MA, USA) was used. The sample (150 μL) was pipetted onto the ATR crystal and measurement was performed at a resolution of 1 cm^{-1} with 32 scans. The air spectrum was measured as a background. The final spectrum was ATR-corrected according to the experimental setup and refractive index of the sample.

High performance liquid chromatography with refractometric detection (HPLC-RI): The method was based on the determination of esters in the side glycerol phase [33], but was significantly modified. The reverse-phase was used: YMC Carotenoid column C30 (250 × 3.0 mm ID with a particle size 3 μm) and the C18 column (150 × 4.6 mm ID with a particle size 7 μm) were connected in series. The methanol (Merc, p.a.) with a flow rate of 0.5 mL/min was used as a mobile phase. The determination was carried out at HPCL (ECOM 2000s, Czech Republic) with a refractometric detector (Shodex RI-101, Shoko Scientific, Yokohama, Japan). The methyl esters were identified by standards and calibration was carried out.

Gas chromatography—mass spectrometry: The Agilent 7890B/5977A Series GC/MSD (Agilent Technologies, Waldbronn, Germany) equipped with autosampler (Agilent 7693) and operating in the electron ionization (EI) mode was used for monitoring and identification of compounds in the reaction mixture. The electron energy of EI was 70 eV, the source temperature was 300 °C, the quadrupole temperature was 150 °C and the transfer line temperature was 300 °C. MS data were acquired over mass range of 50–500 at the rate of 6 scan/s. A TRACE™ TR-FAME capillary column, 60 m × 0.25 mm I.D., film thickness 0.25 μm (Thermo Fisher Scientific, MA, USA) under gradient conditions, was used for separation of all compounds in the sample. Helium at a flow rate of 1 mL/min was used as the carrier gas with the following oven temperature program: 70 °C held for 3.5 min, gradient 90 °C/min to 160 °C held for 2 min, gradient 5 °C/min to 200 °C held to 1 min, and gradient 2 °C/min to 240 °C held to 50 min. An amount of 0.5 μL of the sample was injected to GC under split-mode injection with a 1:50 split ratio at 250 °C. The determination was carried out twice. All samples were prepared as follows: 0.25 g of esters was dissolved in 3 g of acetonitrile, then the mixture was diluted five times with acetonitrile containing the internal standard (final concentration 0.4 mg/mL). The internal standard was C17:0 methyl ester (≥99.0%, Merck KGaA, Darmstadt, Germany). The area of the peaks (A) and thus the representation of individually detected compounds was related to the internal standard (A_{IS}).

Other methods: The esters were characterized by the iodine value (IV) and water content according to EN 14214. Moreover, the epoxide index (EI) (EN 3001) and the simulated distillation [26] were used for determination. The ^{1}H-NMR was determined in deuterated chloroform at 500 MHz (Bruker Ascend) with internal calibration (0.0 ppm).

4. Conclusions

Several methods were applied for the identification of intermediates and products of the epoxidation of methyl esters with variable profiles of higher fatty acids. The composition of esters significantly influenced the course of the epoxidation reaction. The infrared spectroscopy allows to determine only the types of functional group of esters (not each type of ester) including ring-opening reaction (alcohols). For high pressure liquid chromatography, it is possible to separate the esters without and with oxirane rings, but separation of individually epoxides is not possible (only together). The gas chromatography with mass spectrometry was applied for identification of the individual reaction components: esters with different amounts and positions of oxirane rings (together 17 products). Therefore, gas chromatography allows to control the epoxidation process by determination and quantification of intermediates and products. The time course and the selectivity for each reaction component is possible to calculate, which allows to suggest the reaction mechanism, including the calculation of rate constants. The epoxidation will be more environmentally friendly because less by-products will be formed. The epoxidation is suitable for oils crops (especially non-edible ones, such as *Camelina sativa*), which are not possible to be used as a fuel due to high content of unsaturated fatty acids.

Supplementary Materials: The following supporting information can be downloaded at: https://www.mdpi.com/article/10.3390/molecules28062819/s1, Figure S1: ^{1}H NMR for methyl esters and epoxidized methyl esters of *Camelia sativa* oil; Table S1: Retention time (t_r) of fatty acid methyl esters and products of epoxidation.

Author Contributions: Conceptualization, M.H.; methodology, T.H.; validation, T.H.; investigation, T.H., K.F. and D.K.; resources, A.P.; writing—original draft preparation, M.H.; writing—review and editing, M.H. and T.H.; visualization, D.K.; supervision, M.H.; All authors have read and agreed to the published version of the manuscript.

Funding: The work was supported by University of Pardubice, Czech Republic, project SGS_2023_008.

Institutional Review Board Statement: Not applicable.

Informed Consent Statement: Not applicable.

Data Availability Statement: The data presented in this study are available on request from the authors.

Conflicts of Interest: The authors declare no conflict of interest.

Sample Availability: Samples of the compounds are available from the authors.

Nomenclature

E_ME	epoxide methyl ester (methyl ester with oxirane ring)
FA	fatty acids
FTIR	Fourier-transform infrared spectroscopy
GC-MS	gas chromatography with mass spectrometry
HPLC-RI	high performance liquid chromatography with refractometric detection
ME	methyl ester
SO	sunflower oil
LO	linseed oil
RO	rapeseed oil
IV	iodine value (g I_2/100 g)
EI	epoxide index (mol/kg)

References

1. Ben Bacha, A.; Alonazi, M.; Alharbi, M.G.; Horchani, H.; Ben Abdelmalek, I. Biodiesel Production by Single and Mixed Immobilized Lipases Using Waste Cooking Oil. *Molecules* **2022**, *27*, 8736. [CrossRef] [PubMed]
2. Hajek, M.; Vavra, A.; Carmona, H.D.; Kocik, J. The Catalysed Transformation of Vegetable Oils or Animal Fats to Biofuels and Bio-Lubricants: A Review. *Catalysts* **2021**, *11*, 1118. [CrossRef]
3. Tan, H.W.; Aziz, A.R.A.; Aroua, M.K. Glycerol production and its applications as a raw material: A review. *Renew. Sustain. Energy Rev.* **2013**, *27*, 118–127. [CrossRef]
4. Karak, N. *Vegetable Oil-Based Epoxies*; Woodhead Publishing: Sawston, England, 2012; ISBN 9780857097101.
5. Nogales-Delgado, S.; Cabanillas, A.G.; Romero, A.G.; Martin, J.M.E. Monitoring tert-Butylhydroquinone Content and Its Effect on a Biolubricant during Oxidation. *Molecules* **2022**, *27*, 8931. [CrossRef]
6. Atzori, L.; Comes, A.; Fusaro, L.; Aprile, C.; Cutrufello, M.G. Aluminosilicate-Supported Catalysts for the Synthesis of Cyclic Carbonates by Reaction of CO_2 with the Corresponding Epoxides. *Molecules* **2022**, *27*, 8883. [CrossRef]

7. Manka, D.; Siewniak, A. Deep Eutectic Solvents as Catalysts for Cyclic Carbonates Synthesis from CO_2 and Epoxides. *Molecules* **2022**, *27*, 9006. [CrossRef]
8. Dunn, R.O. Thermal analysis of alternative diesel fuels from vegetable oils. *J. Am. Oil Chem. Soc.* **1999**, *76*, 109–115. [CrossRef]
9. Orsavova, J.; Misurcova, L.; Ambrozova, J.; Vicha, R.; Mlcek, J. Fatty Acids Composition of Vegetable Oils and Its Contribution to Dietary Energy Intake and Dependence of Cardiovascular Mortality on Dietary Intake of Fatty Acids. *Int. J. Mol. Sci.* **2015**, *16*, 12871–12890. [CrossRef]
10. Fukuda, H.; Kondo, A.; Noda, H. Biodiesel fuel production by transesterification of oils. *J. Biosci. Bioeng.* **2001**, *92*, 405–416. [CrossRef] [PubMed]
11. Bhuiya, M.M.K.; Rasul, M.G.; Khan, M.M.K.; Ashwath, N. Biodiesel production and characterisation of poppy (*Papaver somniferum* L.) seed oil methyl ester as a source of 2nd generation biodiesel feedstock. *Ind. Crops Prod.* **2020**, *152*, 112493. [CrossRef]
12. Wahl, H.G.; Habel, S.Y.; Schmieder, N.; Liebich, H.M. Identification of Cis-Trans-Isomers of Methyl-Ester and Oxazoline Derivatives of Unsaturated Fatty-Acids Using Gc-Ftir-Ms. *J. High Resolut. Chromatogr.* **1994**, *17*, 543–548. [CrossRef]
13. Santacesaria, E.; Turco, R.; Russo, V.; Tesser, R.; Di Serio, M. Soybean Oil Epoxidation: Kinetics of the Epoxide Ring Opening Reactions. *Processes* **2020**, *8*, 1134. [CrossRef]
14. Kousaalya, A.B.; Beyene, S.D.; Gopal, V.; Ayalew, B.; Pilla, S. Green epoxy synthesized from Perilla frutescens: A study on epoxidation and oxirane cleavage kinetics of high-linolenic oil. *Ind. Crop Prod.* **2018**, *123*, 25–34. [CrossRef]
15. Mecozzi, F.; Dong, J.J.; Angelone, D.; Browne, W.R.; Eisink, N. Oxidative Cleavage of Alkene C=C Bonds Using a Manganese Catalyzed Oxidation with H2O2 Combined with Periodate Oxidation. *Eur. J. Org. Chem.* **2019**, *2019*, 7151–7158. [CrossRef] [PubMed]
16. Nogales-Delgado, S.; Martin, J.M.E.; Ocana, M.S. Use of mild reaction conditions to improve quality parameters and sustainability during biolubricant production. *Biomass Bioenerg.* **2022**, *161*, 106456. [CrossRef]
17. Pantone, V.; Laurenza, A.G.; Annese, C.; Fracassi, F.; Fusco, C.; Nacci, A.; Russo, A.; D'Accolti, L. Methanolysis of epoxidized soybean oil in continuous flow conditions. *Ind. Crop Prod.* **2017**, *109*, 1–7. [CrossRef]
18. Polese, R.; Pintus, E.; Nuvoli, L.; Tiana, M.; Pintus, S.; Satta, G.; Beccu, A.; Gaspa, S.; Carraro, M.; De Luca, L.; et al. Aquivion perfluorosulfonic superacid as an effective catalyst for selective epoxidation of vegetable oils. *R. Soc. Open Sci.* **2022**, *9*, 211554. [CrossRef]
19. Japir, A.; Salimon, J.; Derawi, D.; Bahadi, M.; Yusop, M.R. Separation of Free Fatty Acids from High Free Fatty Acid Crude Palm Oil Using Short-Path Distillation. *AIP Conf. Proc.* **2016**, *1784*, 030001. [CrossRef]
20. Ceron, A.A.; Boas, R.N.V.; Biaggio, F.C.; de Castro, H.F. Synthesis of biolubricant by transesterification of palm kernel oil with simulated fusel oil: Batch and continuous processes. *Biomass Bioenerg.* **2018**, *119*, 166–172. [CrossRef]
21. Mungroo, R.; Pradhan, N.C.; Goud, V.V.; Dalai, A.K. Epoxidation of canola oil with hydrogen peroxide catalyzed by acidic ion exchange resin. *J. Am. Oil Chem. Soc.* **2008**, *85*, 887–896. [CrossRef]
22. La Scala, J.; Wool, R.P. Effect of FA composition on epoxidation kinetics of TAG. *J. Am. Oil Chem. Soc.* **2002**, *79*, 373–378. [CrossRef]
23. Sammaiah, A.; Padmaja, K.V.; Prasad, R.B.N. Synthesis of Epoxy Jatropha Oil and its Evaluation for Lubricant Properties. *J. Oleo Sci.* **2014**, *63*, 637–643. [CrossRef] [PubMed]
24. Xia, W.; Budge, S.M.; Lumsden, M.D. H-1-NMR Characterization of Epoxides Derived from Polyunsaturated Fatty Acids. *J. Am. Oil Chem. Soc.* **2016**, *93*, 467–478. [CrossRef]
25. Piazza, G.J.; Nunez, A.; Foglia, T.A. Epoxidation of fatty acids, fatty methyl esters, and alkenes by immobilized oat seed peroxygenase. *J. Mol. Catal. B Enzym.* **2003**, *21*, 143–151. [CrossRef]
26. Mushtaq, M.; Tan, I.M.; Nadeem, M.; Devi, C.; Lee, S.Y.C.; Sagir, M.; Rashid, U. Epoxidation of methyl esters derived from Jatropha oil: An optimization study. *Grasas Aceites* **2013**, *64*, 103–114. [CrossRef]
27. Bachler, C.; Schober, S.; Mittelbach, M. Simulated Distillation for Biofuel Analysis. *Energy Fuels* **2010**, *24*, 2086–2090. [CrossRef]
28. Yuan, W.; Hansen, A.C.; Zhang, Q. Vapor pressure and normal boiling point predictions for pure methyl esters and biodiesel fuels. *Fuel* **2005**, *84*, 943–950. [CrossRef]
29. Lee, P.L.; Yunus, W.M.Z.W.; Yeong, S.K.; Abdullah, D.K.; Lim, W.H. Optimization of the Epoxidation of Methyl Ester of Palm Fatty Acid Distillate. *J. Oil Palm. Res.* **2009**, *21*, 675–682.
30. Wadumesthrige, K.; Salley, S.O.; Ng, K.Y.S. Effects of partial hydrogenation, epoxidation, and hydroxylation on the fuel properties of fatty acid methyl esters. *Fuel Process. Technol.* **2009**, *90*, 1292–1299. [CrossRef]
31. Turco, R.; Tesser, R.; Russo, V.; Cogliano, T.; Di Serio, M.; Santacesaria, E. Epoxidation of Linseed Oil by Performic Acid Produced In Situ. *Ind. Eng. Chem. Res.* **2021**, *60*, 16607–16618. [CrossRef]
32. de Haro, J.C.; Izarra, I.; Rodriguez, J.F.; Perez, A.; Carmona, M. Modelling the epoxidation reaction of grape seed oil by peracetic acid. *J. Clean. Prod.* **2016**, *138*, 70–76. [CrossRef]
33. Hajek, M.; Skopal, F.; Kwiecien, J.; Cernoch, M. Determination of esters in glycerol phase after transesterification of vegetable oil. *Talanta* **2010**, *82*, 283–285. [CrossRef]
34. Holcapek, M.; Lisa, M.; Jandera, P.; Kabatova, N. Quantitation of triacylglycerols in plant oils using HPLC with APCI-MS, evaporative light-scattering, and UV detection. *J. Sep. Sci.* **2005**, *28*, 1315–1333. [CrossRef] [PubMed]
35. Anuar, S.T.; Zhao, Y.Y.; Mugo, S.M.; Curtis, J.M. Monitoring the Epoxidation of Canola Oil by Non-aqueous Reversed Phase Liquid Chromatography/Mass Spectrometry for Process Optimization and Control. *J. Am. Oil Chem. Soc.* **2012**, *89*, 1951–1960. [CrossRef]

36. Christie, W.W. Mass Spectrometry of Methyl Esters. Available online: https://www.lipidmaps.org/resources/lipidweb/lipidweb_html/ms/methesters/me-epoxy/index.htm (accessed on 10 January 2023).
37. Vaclavik, A.; Christian, E.W. *Essentials of Food Science*; Springer Science & Business Media: Berlin/Heidelberg, Germany, 2007.
38. Joback, K.G.; Reid, R.C. Estimation of Pure-Component Properties from Group-Contributions. *Chem. Eng. Commun.* **1987**, *57*, 233–243. [CrossRef]
39. Akasaka, K.; Akama, T.; Ohrui, H.; Meguro, H. Measurement of Hydroxy and Hydroperoxy Fatty-Acids by a High-Pressure Liquid-Chromatography with a Column-Switching System. *Biosci. Biotechnol. Biochem.* **1993**, *57*, 2016–2019. [CrossRef]
40. Hajek, M.; Skopal, F.; Machek, J. Simplification of separation of the reaction mixture after transesterification of vegetable oil. *Eur. J. Lipid Sci. Technol.* **2008**, *110*, 347–350. [CrossRef]
41. Musil, M.; Hajek, M.; Skopal, F.; Vavra, A. Improved method of water removal from vegetable oil. *Chem. Pap.* **2019**, *73*, 767–769. [CrossRef]

molecules

MDPI

Article

Evidence for the Hydration of Some Organic Compounds during Reverse-Phase HPLC Analysis

Igor G. Zenkevich *, Abdennour Derouiche and Daria A. Nikitina

Institute for Chemistry, St. Petersburg State University, 198504 St. Petersburg, Russia
* Correspondence: izenkevich@yandex.ru; Tel.: +7-(812)-428-4045

Abstract: Some polar analytes (X) can reversibly form hydrates in water-containing eluents under the conditions of reversed-phase HPLC analysis, $X + H_2O \rightleftarrows X \times H_2O$. One of the methods to detect their formation is the recurrent approximation of the net retention times of such analytes, $t_R(C + \Delta C) = at_R(C) + b$, where ΔC = const is the constant step in the variation of the organic modifier content of an eluent. These dependencies are linear if hydrates are not formed, but in the case of hydrate formation, they deviate from linearity under high water content. It has been shown that UV spectroscopic parameters, namely, relative optical densities: $A_{rel} = A(\lambda_1)/A(\lambda_2)$, depend on eluent composition for some organic compounds, but their variations cannot be used as indicators for hydrate formation. The coefficients that characterize the dependence of the analyte retention indices on the organic component concentration of an eluent, dRI/dC, appeared to be the most informative additional criterion for hydration. The values of these coefficients for most polar analytes are largely negative ($dRI/dC < 0$), whereas, for nonpolar compounds, they are largely positive ($dRI/dC > 0$).

Keywords: reverse-phase HPLC; hydration of analytes; recurrent approximation of retention times; retention indices; dependence of indices on the concentration of an organic modifier in an eluent

1. Introduction

The principal advantage of high-performance liquid chromatography (HPLC) compared to gas chromatography is its applicability to nonvolatile and thermally unstable analytes [1]. At the same time, the main disadvantage of reversed-phase (RP) HPLC is the risk of the hydrolysis of some analytes due to the presence of water in the eluent. The "intermediate option", which is rarely taken into account, is the reversible formation of the hydrates of some analytes during their chromatographic separation in water-containing eluents.

The formation of hydrates is the typical property of numerous inorganic compounds [2]. Most such hydrates are stable and can be isolated in a solid state. However, unexpectedly, many organic compounds (X) also form hydrated forms, preferably monohydrates. Instead of the expression "formation of hydrates", the following equilibrium seems to be more rigorous:

$$X + H_2O \rightleftarrows X \times H_2O \quad (1)$$

The probability of hydrate formation is determined by the constant of hydration, K_{hydr}:

$$K_{hydr} = [X \times H_2O]/\{[X] \times [H_2O]\} \quad (2)$$

If $K_{hydr} << 1$, the formation of hydrates in an aqueous media can be neglected, but the inequality $K_{hydr} >> 1$ corresponds to relatively stable hydrates. Some of them can be isolated so that their physicochemical properties can be experimentally determined. CAS numbers are assigned to numerous hydrates, both stable and unstable. Several examples of hydrates of both kinds are presented in Table 1. Some organic compounds form stable covalent hydrates (e.g., trifluoroacetaldehyde, hexafluoroacetone, ninhydrin, etc.).

Table 1. Some examples of reference data for hydrates of organic compounds.

Compound	CAS No. (Anhydrous Form)	CAS No. (Hydrate)	Composition and Properties * (if Known)
Unstable hydrates			
Methanol	67-56-1	118240-86-1 151900-28-5	1:1
Acetonitrile	75-05-8	128870-13-3	1:1
Acetic acid	64-19-7	19215-29-3 99294-94-7	1:1, 1:2
Anthracene	120-12-7	188974-01-8	1:1
Stable noncovalent hydrates			
Ethylene diamine	107-15-3	6780-13-8	1:1; T_b 118; n_D^{20} 1.448-1.451; d_4^{20} 0.96
Citric acid	77-92-9	5949-29-1	1:1 **
Caffeine	58-08-2	5743-12-4	1:1 **
Benzene-1,2,3-tricarboxylic (hemimellitic) acid	569-51-7	732304-21-1 (mono); 36362-97-7 (di)	1:1; T_m 190-192

(*) Abbreviations: T_b—normal boiling point, T_m—melting point, n_D^{20}—index of refraction, d_4^{20}—relative density; (**) hydrates decompose below the melting point.

The information on hydrates in this table is taken both from original publications and (mostly) from the webpages of chemical companies (more detailed collections of the data for the hydrates are presented in [3,4]).

Because hydrates are compounds that are definitely more polar than the anhydrous forms of organic compounds, their formation may account for some anomalies of their retention in RP HPLC, depending on the ratio between water and the organic modifier in an eluent. If $K_{hydr} << 1$, the eluent contains solely the nonhydrated form of the analyte and there should be no anomalies of its retention under any eluent composition. On the other hand, if $K_{hydr} >> 1$, we can assume the predominance of the hydrated form of the analyte with which no transformations take place with variation in the eluent composition; hence, no retention anomalies are observed as well. The most interesting case is the comparable content of the nonhydrated and hydrated forms of analytes in an eluent (both forms coexist together), i.e., when $K_{hydr} \approx 1$. In this case, the variations of the ratio of the organic and aqueous phases in an eluent should strongly influence the position of the equilibrium (2) and the ratio of the nonhydrated and hydrated forms, causing unpredictable variations in the retention parameters of such analytes. Numerous equations for approximation of the dependencies of the retention times on the content of organic modifiers have been proposed (see, e.g., [5,6]). It is important that most of them become inapplicable if the analytes reversibly form hydrates in an eluent. Actually, hydrate formation is the chemical transformation of an analyte during chromatographic separation depending on the organic modifier concentration.

Thus, the problem of the HPLC detection of reversibly formed hydrates is that the retention times correspond not to the sole structures but to at least two of the different forms of the analytes in variable proportions, depending on the eluent composition. Detecting relatively small anomalies in the t_R-values against the background of their significant variations due to the dependence $t_R = f(C)$ seems to be rather difficult. Let us briefly discuss the possible effects of hydrate formation using easily perceived examples.

The unusual anomalies of chromatographic retention caused by the formation of hydrates were revealed for the first time for several complex polyfunctional synthetic antitumor drugs produced by Biokad JSC (St. Petersburg, Russia) [7]. Because the direct presentation of the t_R values as a function of the concentration of the organic modifier in an eluent (*C*) does not reveal most of the anomalies, the so-called recurrent representation of the retention times was used:

$$t_R(C + \Delta C) = a t_R(C) + b, \quad (3)$$

where ΔC is the constant increment in the variations of the organic modifier concentration in an eluent, and the coefficients *a* and *b* are calculated by the least squares method (LSM).

A short description of the properties of the recurrent relations is discussed below (Section 3.1). Here, it seems important to compare the plots of the recurrent dependencies (3) for three drugs with the trivial names gefitinib (Figure 1, structure I), pazopanib (II), and imatinib (III):

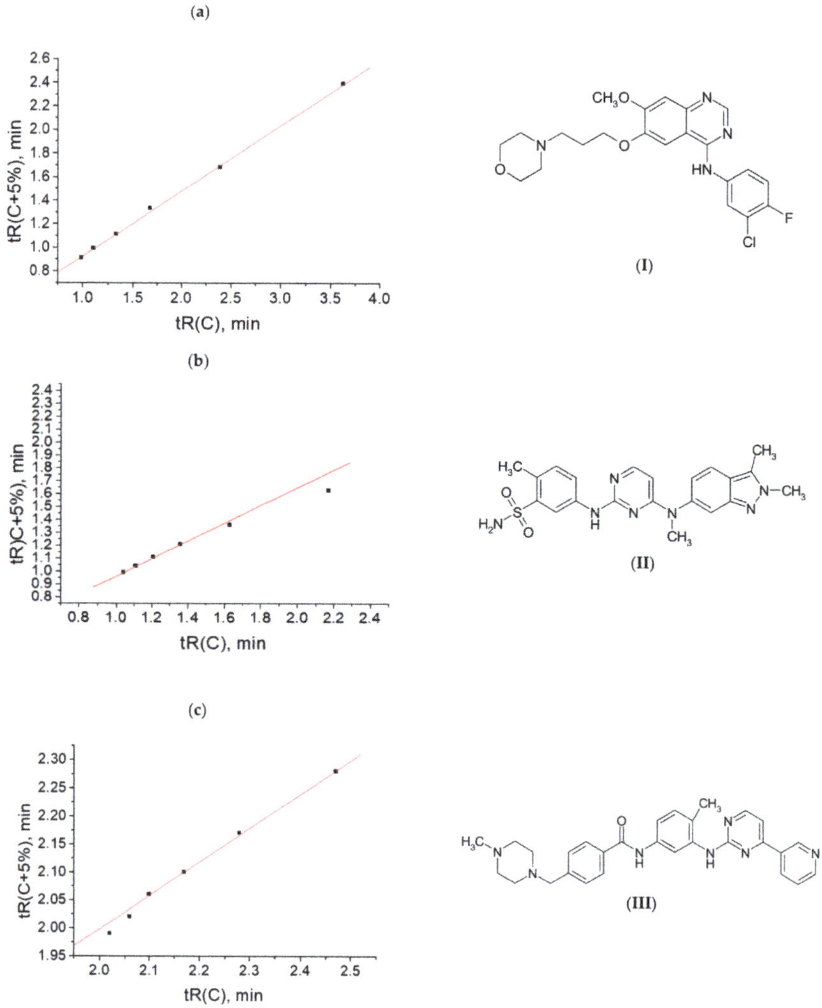

Figure 1. Examples of some of the typical features in the recurrent approximation plots of the retention times of (**a**) gefitinib (**I**), (**b**) pazopanib (**II**), and (**c**) imatinib (**III**); all data were obtained with acetonitrile–water eluents. See text for detailed comments.

All the plots are based on the raw retention times within acetonitrile concentration ranges of 35–65% v/v (compounds I and III) and 20–50% v/v (compound II), varied with 5% steps (seven experimental t_R values for each analyte). This gives six points for the recurrent dependencies, which lie on the same straight line for gefitinib (I) (correlation coefficient $R = 0.9996$). On the contrary, the corresponding plot for pazopanib (II) has a linear section (four "left" points) with $R = 0.9998$ but two "right" points that deviate (down) from the regression line. The right parts of all plots (largest argument values) correspond to the eluents that contain the largest amounts of water when the equilibria (1) are shifted toward the formation of more hydrophilic hydrates. Other deviations in the regression data approximations can be observed for imatinib (III): the four right points fall on a straight line ($R = 0.9998$), whereas the two left points (for the eluents with the highest acetonitrile content) deviate down from the regression line. This means that the recurrent approximation is a sensitive tool for revealing "fine" anomalies in retention data.

This information on hydrate formation seems to be rather important for different analytical applications. For example, the values of the so-called hydrophobicity factor (log$P$) are considered to represent the valuable characteristics of organic compounds, including drugs. Different kinds of software (e.g., ACD, ChemAxon, etc.) are recommended for the theoretical evaluation of these parameters. However, all such calculations can be made for nonhydrated molecules. If the target analyte forms a hydrate, the precalculated log$P$ values are very different from the experimental values.

Because the detection of hydrate formation in chromatographic eluents appeared to be a difficult task (in particular), and the possibilities of the recurrent approximation of the retention data in RP HPLC require additional characterization (in general), we consider these problems in our paper.

2. Results

2.1. Measuring the Retention Times of Selected Analytes

Up to now, revealing the dependencies of the retention parameters (t_R) on eluent composition (usually on the content of the organic modifier, C) remains the main trick in the HP HPLC characterization of various organic compounds on different sorbents (see, e.g., [8–11]). A few dozen different equations have been proposed for approximating this dependence $t_R(C)$ [6]. Our task was to characterize the features of those compounds forming the hydrates in eluents. For this purpose, we have selected about 30 model compounds for the analyses with the methanol–water eluents and about 20 compounds for the analyses with the acetonitrile–water eluents. The t_R measurement is a standard procedure that does not require special detailed description. It is unnecessary to consider all the numerical $t_R(C)$ data. The retention times of the selected analytes were measured within the ranges 50–85% vol. methanol and 35–70% vol. acetonitrile, with 5% concentration steps.

The important detail of the selection of the model analytes was to avoid the coincidence of their pK_a values with the pH of the eluents. For the acetonitrile-containing eluents without acidic or salt constituents, the reference pH value was approximately 5.6, whereas for the eluent 1:1 v/v methanol + 0.1% trifluoroacetic acid, the pH was 2.7. The pK_a values of some of the analytes are as follows: 1H-benzotriazole 8.5 ± 0.1, phthalimide: 8.2 or 10.2, 1-phenylpyrazolidin-3-one: 7.5 and 9.5, diethyl-*m*-toluamide: −1.37, all N-substituted *p*-toluenesulfonamides: 11.2 ± 0.8 (average value for 10 compounds), sulfamethoxazole: 5.7 ± 0.3, sulfamerazine: 7.0, and *p*-toluilic acid: 4.4. Due to the pK_a value, the latter acid was excluded from further consideration. The most "suspicious" value in the above series is 5.7 ± 0.3 for sulfamethoxazole (close to the 5.6 pH of the acetonitrile-containing eluents), but in the aqueous solution, this compound exists as hydrate, with a different pK_a value.

Even small uncontrolled variations in the eluent flow rate can affect the experimental results [12,13]. This is manifested to the greatest extent in the water–methanol eluents because the viscosity of the CH_3OH–water mixtures is maximized under an approximate 40 vol.% methanol content. If the HPLC pump(s) does not provide the fixed eluent flow rate under increasing eluent viscosity, this may lead to unpredictable distortions of the t_R-values. This is why the use of one of the available HPLC instruments to us was rejected [14].

2.2. Calculation of Retention Indices

As all the analyses of selected compounds were carried under isocratic conditions, all of them were characterized by logarithmic (Kovats) retention indices [15]:

$$RI_x = RI_n + (RI_{n+1} - RI_n) \times [\log(t_{R,x}') - \log(t_{R,n}')]/[\log(t_{R,n+1}') - \log(t_{R,n}')] \quad (4)$$

where $t_{R,x}$, $t_{R,n}$, and $t_{R,n+1}$ are the net retention times of the target analyte and the two reference compounds eluted immediately before and immediately after (*n*-alkyl phenyl ketones), and RI_x, RI_n, and RI_{n+1} are their retention indices and the prime means conversion of net retention times to the adjusted retention times, $t_R' = t_R - t_0$, where t_0 is the retention time of the theoretically unabsorbed component ("dead time").

The required t_0-values were calculated using the t_R values for the three serial homologs of the *n*-alkyl phenyl ketones using the Peterson and Hirsch relationship [16]:

$$t_0 = (t_{R,1}t_{R,3} - t_{R,2}^2)/(t_{R,1} + t_{R,3} - 2t_{R,2}) \quad (5)$$

Relation (4) is equivalent to the following linear dependence (coefficients *a* and *b* are calculated by LSM):

$$RI_x = a\log(t_{R,x}') + b \quad (6)$$

This means that the calculation of the retention indices is possible not only by interpolation ("between" reference compounds) but, in some cases, by extrapolation (out of the range of the retention times of the reference compounds).

The retention indices of some of the organic compounds determined using methanol as the organic component of an eluent are listed in Table 2; the data for the acetonitrile-containing eluents are presented in Table 3.

Table 2. Retention indices of some organic compounds, depending on the methanol content in the eluent.

Analyte	MW	N{H}	Methanol Content (% *v*/*v*)							
			50	55	60	65	70	75	80	85
Toluene	92	0	1052	1071	1088	1107	1127	1146	1176	-
o-Xylene	106	0	1150	1166	1186	1202	1230	1254	1292	-
Chlorobenzene	112	0	1046	1057	1067	1078	1090	1105	1127	1144
1H-Benzotriazol	119	1	688	686	684	684	684	680	681	-
Acetophenone	120	0	800	800	800	800	800	800	800	800
4-Methylbenzaldehyde	120	0	870	-	874	-	879	-	884	-
2-Hydroxybenzaldehyde	122	1	795	-	804	-	811	-	819	-
Nitrobenzene	123	0	847	849	854	856	860	857	852	-
Acetophenone hydrazone	134	2	-	738	737	738	734	737	736	734
4-Methylbenzaldehyde hydrazone	134	2	-	719	720	720	721	724	724	722
2,3,5-Trimethylphenol	136	1	852	854	858	862	862	864	860	-
2-Hydroxybenzaldehyde hydrazone	136	3		722	719	715	713	711	708	702
3-Nitrophenol	139	1	799	796	793	791	784	776	768	-
Phthalimide	147	1	693	692	691	689	690	686	686	-

Table 2. *Cont.*

Analyte	MW	*N*{H}	Methanol Content (% *v*/*v*)							
			50	55	60	65	70	75	80	85
4-Methylacetophenone hydrazone	148	2	-	784	799	803	830	-	-	-
Butyrophenone hydrazone	162	2	-	875	885	894	901	890	905	939
1-Phenylpyrazolidin-3-one	162	1	730	714	692	670	732	729	715	-
Ninhydrine (hydrate)	178	2	662	664	664	666	668	664	663	-
N-Allyl-*p*-toluenesulfonamide	211	1	852	838	823	808	792	772	756	732
N,N-Diethyl-*p*-toluenesulfonamide	227	0	978	964	950	936	920	903	885	862
N-*tert*-Butyl-*p*-toluenesulfonamide	227	1	968	952	935	918	898	876	852	824
N-Phenyl-*p*-toluenesulfonamide	247	1	963	942	918	895	869	842	813	782
N-Hexyl-*p*-toluenesulfonamide	255	1	1225	1205	1185	1165	1140	1110	1075	1029
N-Benzyl-*p*-toluenesulfonamide	261	1	1014	993	972	948	921	894	860	828

Table 3. Retention indices of some organic compounds depending on the acetonitrile content in the eluent.

Analyte	MW	Acetonitrile Content (% *v*/*v*)							
		35	40	45	50	55	60	65	70
Toluene	92	1028	1022	1024	1036	1036	1030	1048	-
o-Xylene	106	-	-	1119	1126	1122	1124	1151	-
Chlorobenzene	112	-	1035	1024	1036	1040	1036	1055	-
1H-Benzotriazol	119	694	668	652	654	644	630	654	-
Acetophenone	120	800	800	800	800	800	800	800	-
Nitrobenzene	123	863	864	858	865	860	845	851	-
2,3,5-Trimethylphenol	136	935	925	914	916	905	888	893	-
3-Nitrophenol	139	777	769	755	755	745	724		-
Phthalimide	147	711	692	679	685	676	666	686	-
1-Phenylpyrazolidin-3-one	162	697	671	654	654	642	-	-	-
Diethyl-*m*-toluamide	191	874	865	850	854	846	831	843	-
N-Allyl-*p*-toluenesulfonamide	211	879	870	861	848	837	827	814	816
N,N-Diethyl-*p*-toluenesulfonamide	227	-	1030	1021	1012	1006	997	987	974
N-*tert*-Butyl-*p*-toluenesulfonamide	227	972	962	951	939	930	921	914	903
N-Phenyl-*p*-toluenesulfonamide	247	1010	1005	970	948	928	904	883	-
Sulfamethoxazole (hydrate)	253	717	714	698	699	689	-	-	-
N-Hexyl-*p*-toluenesulfinamide	255	-	-	1208	1184	1174	1157	1139	1129
N-Benzyl-*p*-toluenesulfonamide	261	-	1025	1006	984	967	950	928	-
Sulfamerazine	264	694	668	652	651	638	621	-	-

The symbol $N\{H\}$ in Table 2 and below means the total number of the so-called active hydrogen atoms in a molecule (the number of atoms capable of exchanging with the hydrogen atoms of a solvent). The intra- and interday reproducibility of the RI values in these tables is approximately 1–3 index units (i.u.).

2.3. Evaluation of the Relative Optical Densities

The detection of the hydrates of the analytes ($X \times H_2O$) formed in an eluent can be achieved, at least theoretically, by recording and interpreting the changes in their UV spectra. However, the registration of the absolute UV spectroscopic parameters in HPLC is not reliable enough; hence, the determination of the so-called relative optical densities (A_{rel}) seems to be preferable:

$$A_{rel} = A(\lambda_1)/A(\lambda_2) \approx S(\lambda_1)/S(\lambda_2) \quad (7)$$

where $S(\lambda_1)$ and $S(\lambda_2)$ are the areas of the same chromatographic peak at different wavelengths.

The relative optical densities were recommended as an additional criterion for the identification of the analytes using RP HPLC in combination with the chromatographic parameters [17–19], including the level of the so-called group identification (attribution to the corresponding homologous series with the same chromophores). Table 4 contains the A_{rel} values for some of the organic compounds measured with the methanol–water eluents (the range of the methanol content is 50–85% *v*/*v*), and Table 5 contains the analogous data for the acetonitrile–water eluents (55–70% *v*/*v* acetonitrile content).

Table 4. Relative optical densities $A(254)/A(220)$ of some organic compounds, depending on the methanol content in the eluent.

Analyte	**MW**	**Methanol Content (% *v*/*v*)**							
		50	**55**	**60**	**65**	**70**	**75**	**80**	**85**
Toluene	92	0.14	0.21	0.24	0.26	0.30	0.34	0.37	-
o-Xylene	106	0.08	0.08	0.09	0.10	0.11	0.13	0.14	-
Chlorobenzene	112	0.032	0.031	0.031	0.031	0.031	0.031	0.032	0.028
1H-Benzotriazol	119	3.4	3.0	2.4	2.3	4.2	4.9	6.0	-
Acetophenone	120	3.5	3.5	3.0	3.3	3.1	3.0	2.8	2.8
Nitrobenzene	123	1.5	1.6	1.6	1.6	1.6	1.7	1.8	-
Propiophenone	134	2.9	2.9	2.8	2.7	2.6	2.2	2.0	2.3
Acetophenone hydrazone	134	-	1.23	1.23	1.22	1.20	1.13	1.04	-
2-Methylbenzaldehyde hydrazone	134	-	0.63	0.63	0.59	0.62	0.60	0.53	0.63
4-Methylbenzaldehyde hydrazone	134	1.04	0.94	0.95	1.09	0.99	0.93	0.91	-
p-Toluilic acid	136	0.77	0.74	0.76	0.72	0.81	0.81	0.83	-
2-Hydroxybenzaldehyde hydrazone	136	0.48	0.47	0.45	0.46	0.41	0.40	0.40	-
3-Nitrophenol	139	0.39	0.33	0.38	0.40	0.41	0.45	0.45	-
Phthalimide	147	0.02	0.02	0.02	0.02	0.02	0.02	0.02	-
Butyrophenone	148	2.8	3.0	3.0	2.9	2.8	2.7	2.6	2.3
4-Methylacetophenone hydrazone	148	1.18	1.22	1.18	1.16	1.21	1.19	1.12	-
Propiophenone hydrazone	148	1.33	1.16	1.13	1.03	1.10	1.07	1.09	-
Butyrophenone hydrazone	162	1.30	1.13	1.12	1.08	1.10	1.07	1.01	-
1-Phenylpyrazolidin-3-one	162	2.6	2.7	1.3	1.6	2.2	1.7	2.3	-
Ninhydrine (hydrate)	178	-	0.42	0.48	0.46	0.47	0.44	0.44	-
Diethyl-*m*-toluamide	191	0.13	0.13	0.12	0.12	0.11	0.11	0.12	-

Table 4. *Cont.*

Analyte	MW	Methanol Content (% *v*/*v*)							
		50	55	60	65	70	75	80	85
N-Allyl-*p*-toluenesulfonamide	211	0.063	0.062	0.061	0.059	0.059	0.058	0.057	0.052
N,N-Diethyl-*p*-toluenesulfonamide	227	0.28	0.28	0.28	0.27	0.21	0.26	0.24	0.26
N-*tert*-Butyl-*p*-toluenesulfonamide	227	0.091	0.091	0.087	0.092	0.092	0.093	0.094	0.095
N-Phenyl-*p*-toluenesulfonamide	247	0.26	0.26	0.26	0.26	0.25	0.29	0.27	0.25
N-Hexyl-*p*-toluenesulfonamide	255	0.07	0.069	0.067	0.066	0.066	0.064	0.062	0.064
N-Benzyl-*p*-toluenesulfonamide	261	0.078	0.075	0.074	0.072	0.074	0.065	0.068	0.066

Table 5. Relative optical densities $A(254)/A(220)$ of some organic compounds, depending on the acetonitrile content in the eluent.

Analyte	MW	Acetonitrile Content (% *v*/*v*)						
		35	40	45	50	55	60	65
Toluene	92	0.133	0.133	0.128	0.131	0.132	0.134	0.135
o-Xylene	106	-	0.064	0.063	0.063	0.063	0.063	0.064
Chlorobenzene	112	-	-	0.057	0.063	0.070	0.074	0.078
1H-Benzotriazol	119	3.6	3.9	3.8	3.9	4.1	4.2	4.1
Acetophenone	120	4.1	3.8	3.4	3.4	3.0	2.9	2.7
Nitrobenzene	123	1.6	1.6	1.7	1.8	1.8	1.8	1.7
Propiophenone	134	3.2	3.0	2.7	2.6	2.4	2.2	2.2
2,3,5-Trimethylphenol	136	0.083	0.082	0.081	0.084	0.079	0.075	0.075
3-Nitrophenol	139	0.37	0.39	0.36	0.34	0.439	0.387	-
Phthalimide	147	0.031	0.033	0.035	0.037	0.039	0.041	0.042
Butyrophenone	148	2.4	2.3	2.2	2.1	2.0	1.9	1.9
1-Phenylpyrazolidin-3-one	162	1.39	1.40	1.43	1.44	1.45	-	-
Diethyl-*m*-toluamide	191	0.14	0.15	0.15	0.15	0.15	0.15	0.15
N-Allyl-*p*-toluenesulfonamide	211	0.076	0.073	0.071	0.074	0.067	0.074	0.065
N,N-Diethyl-*p*-toluenesulfonamide	227	-	0.295	0.293	0.288	0.290	0.290	0.290
N-*tert*-Butyl-*p*-toluenesulfonamide	227	0.113	0.110	0.108	0.107	0.104	0.102	0.106
N-Phenyl-*p*-toluenesulfonamide	247	0.31	0.29	0.29	0.28	0.28	0.30	0.30
N-Hexyl-*p*-toluenesulfonamide	255	-	-	0.077	0.074	0.073	0.072	0.071
N-Benzyl-*p*-toluenesulfonamide	261	-	0.16	0.16	0.15	0.16	0.15	0.15
Sulfamerazine	264	0.97	0.86	0.76	0.72	0.64	0.53	0.48

Both tables contain examples of compounds with both ascending and descending dependencies $A_{rel}(C)$, as well as with almost no clearly pronounced dependencies. For instance, the aromatic hydrocarbons (toluene, *o*-xylene) in the methanol–water eluents demonstrate the ascending dependence $A_{rel}(C)$, while in the acetonitrile–water eluents, it slightly descends. The reference compounds in RP HPLC, *n*-alkyl phenyl ketones, are characterized by $dA_{rel}/dC < 0$ in all eluents. The most interesting objects, the N-substituted *p*-toluenesulfonamides, demonstrate practically no dependence regarding their relative optical densities on eluent composition.

The joint consideration of the data in Tables 4 and 5 allows for the following conclusions: (1) the A_{rel} values for the analytes in the methanol–water and acetonitrile–water eluents are not usually equal to each other; (2) in some cases, these values depend on the eluent composition, and (3) the variations of these parameters, depending on the organic modifier concentration, are not directly related to hydrate formation. Despite the negative character of this conclusion, it seems rather important because it prevents further attempts to use spectral parameters for detecting the formation of hydrates.

3. Discussion

3.1. *Recurrent Approximation of Chemical Variables: Important Features*

The simple first-order linear recurrent regressions can be applied to the monotonic functions (A) of the integer (n) (Equation (8)) or the equidistant values of the argument (Δx) (Equation (9)). The first kind of recurrence is applicable to the approximation of the various physicochemical properties of the homologs (functions of the number of carbon atoms in molecules), with the number of carbon atoms being an argument by definition [20]. The second kind of recurrence allows for their application to the functions of the temperature or pressure of chemical systems, as well as the concentrations of their constituents. In the latter case, the steps of the variation of the arguments, Δx, should be fixed:

$$A(n + \Delta n) = aA(n) + b \tag{8}$$

$$B(x + \Delta x) = aB(x) + b, \Delta x = \text{const} \tag{9}$$

Specifically, the latter type of relationship can be used in the approximation of the chromatographic retention parameters as functions of temperature (gas chromatography) or of the organic modifier content of eluent in RP HPLC [21].

Recurrent relationships have several unusual mathematical properties. First, their mathematical equivalent (e.g., for Equation (8)) is the polynomial of the variable degree:

$$A(n) = ka^{\mathrm{n}} + b(a^{\mathrm{n}} - 1)/(a - 1) \tag{10}$$

Hence, recurrence relationships unite the properties of the arithmetic (at $a \equiv 1$ and $b \neq 0$) and geometric (at $0 < a \neq 1$ and $b \equiv 0$) progressions. This fact accounts for their unique approximating "ability", especially for the various properties of the homologs within a homologous series because the number of carbon atoms in the molecule cannot be a noninteger argument by definition. Examples of the applicability of these recurrences to the equidistant values of pressure, temperature, or the concentrations of the constituents are the dependencies of t_R on the temperature in gas chromatography and on the organic modifier content of an eluent in RP HPLC [21]. It is noteworthy that using the recurrent relationships in both gas chromatography and HPLC does not require the preliminary determination or calculation of the so-called "dead" time (t_0). Another feature that seems to be important for plotting the recurrent dependencies is that the values of the arguments are not represented in such plots; every point is fixed by the two "neighboring" values of the functions.

When applied to the retention parameters in RP HPLC, recurrence relationships (9) are most often characterized by the correlation coefficients $R > 0.999$ for those analytes that show no anomalies in their chemical nature (e.g., not involved in prototropic equilibria and form no hydrates or tautomers). However, if two (or more) forms of analytes are present in an eluent (e.g., when $K_{hydr} \approx 1$), deviations from the linearity of the recurrent dependencies (3) can be expected; this is due to the fact that the approximation "ability" of the recurrence is significant but not infinite; any changes in analyte speciation lead to distortions in the linearity of the recurrent dependencies.

Thus, the detection of the reversible formation of the hydrates of organic compounds in aqueous media (including HPLC eluents) should be based on the detailed consideration of the dependencies, $t_R(C)$. The first approach seems to be just revealing the deviations of the recurrent approximation of the net retention times from the linearity under high-water-content eluent. The second approach, which is discussed in this manuscript, considers the features of the retention indices of the analytes in HPLC.

3.2. Numerical Modeling of the Anomalies in the Recurrent Approximation of Retention Times

The application of recurrent approximation for revealing the formation of the hydrates of analytes can be illustrated by the following numerical example.

If an analyte (X) forms a hydrate ($X \times H_2O$) in an eluent, then its retention time can be expressed (very roughly, without considering the details) as an arithmetic mean of the retention times of the nonhydrated and hydrated forms:

$$t_R \approx [t_R(X) + t_R(X \times H_2O)]/2 \tag{11}$$

It is logical to believe that the hydrated form of an analyte is more hydrophilic than the nonhydrated form; hence, $t_R(X \times H_2O) < t_R(X)$, or $t_R(X \times H_2O)(C - y) \approx t_R(X)(C)$.

In order to simplify our numerical model, let us assume that the retention factor (k) is inversely proportional to the volume fraction of the organic component of an eluent (x). Such prerequisites correspond to the Row model [22]:

$$1/k = ax + b \tag{12}$$

Moreover, let us postulate that $a = 1$ and $b = 0$, using the net retention times instead of the k-values, which gives $t_R \sim 1/x$. Let us also imagine that the content of the organic modifier in an eluent varies from 0.4 to 1.0 with a step of 0.05. This gives the following set of t_R-values:

$0.4 \leq x \leq 1.0$	0.4	0.45	0.5	0.55	0.6	0.65	0.7	0.75	0.8	0.85	0.9	0.95	1.0
t_R	2.50	2.22	2.00	1.82	1.67	1.54	1.43	1.34	1.25	1.18	1.11	1.05	1.00

The plot of this dependence $t_R(x)$ is the plot of a hyperbolic function (Figure 2a).

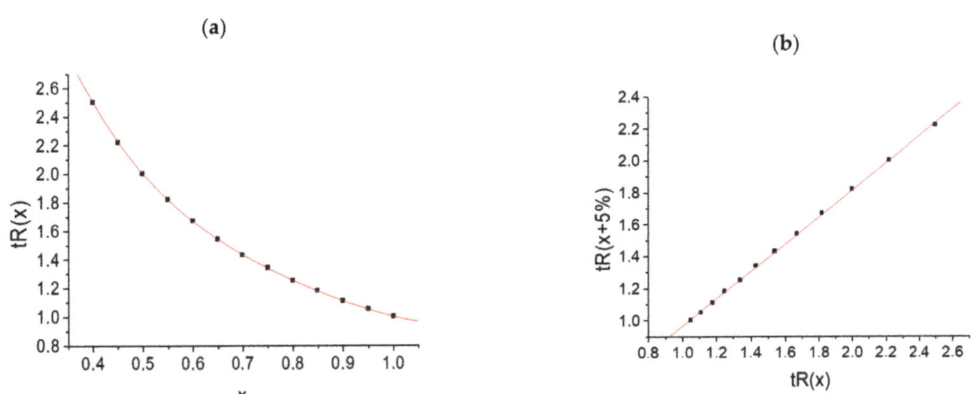

Figure 2. (**a**) Typical nonlinear dependence $t_R(x)$ (numerical modeling) and (**b**) linear recurrent approximation for the same set of data.

The recurrent approximation (Equation (3)) of the same data set with $\Delta C = 5\%$ steps gives a linear regression with the following parameters: $a = 0.847 \pm 0.006$, $b = 0.118 \pm 0.010$, $R = 0.9998$, and $S_0 = 0.009$; the plot of this dependence is shown in Figure 2b. It looks typical for analytes forming no hydrates in eluents (having no anomalies of chromatographic retention).

Let us assume that analyte X forms more hydrophilic hydrate $X \times H_2O$ and that the retention time of this hydrate approximately corresponds to that of the parent compound X at the higher content of the organic modifier, $t_R(X \times H_2O)(C) \approx t_R(X)(C + y)$. If we accept $y = 0.2$, we obtain the following set of numerical estimations and, finally, the target retention times, t_R* (the last line below):

$0.2 \leq x \leq 0.8$	0.2	0.25	0.3	0.35	0.4	0.45	0.5	0.55	0.6	0.65	0.7	0.75	0.8
t_R(non-hydrate)	5.00	4.00	3.33	2.86	2.50	2.22	2.00	1.82	1.67	1.54	1.43	1.34	1.25
$0.4 \leq x \leq 1.0$	0.4	0.45	0.5	0.55	0.6	0.65	0.7	0.75	0.8	0.85	0.9	0.95	1.0
t_R(hydrate)	2.50	2.22	2.00	1.82	1.67	1.54	1.43	1.34	1.25	1.18	1.11	1.05	1.00
$t_R^* = [t_R + t_R(\text{hydrate})]/2$	3.75	3.11	2.66	2.34	2.08	1.88	1.72	1.58	1.46	1.36	1.37	1.19	1.12

Surprisingly, the recurrent approximation of the set of t_R*-values (the sum of two hyperbolic functions) in comparison with the plot in Figure 2b visually demonstrates the detected deviations from linearity (Figure 3) only in the area of the large t_R*-values, corresponding to the high water content of an eluent.

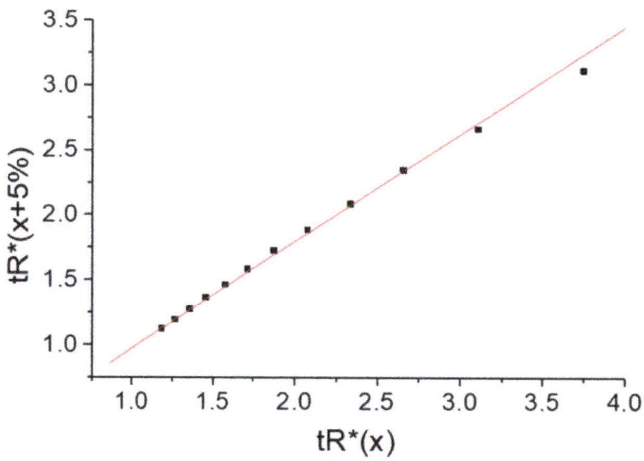

Figure 3. Plot of the recurrent approximation of the superposition of two nonlinear hyperbolical dependencies $t_R(x)$ (numerical modeling). The area of maximal t_R*-values shows visible deviations from linearity.

Such features of the recurrent approximation plots are typical of analytes forming hydrates in eluents.

3.3. Revealing Those Compounds That Are Reversibly Forming Hydrates

Hydrate formation can be readily confirmed for solid substances (both inorganic and organic). This can be carried out using differential scanning thermogravimetry or even "classical" elemental analysis. The detection of the unstable hydrates of organic compounds in solutions (when their isolation is impossible) is much more difficult. In some cases, hydrate formation can be inferred from the appearance of new bands in the IR and UV spectra. Mass spectrometric methods provide no information on the formation of hydrates in solutions.

As mentioned above, some polyfunctional synthetic drugs were the first examples of the application of recurrent approximation of net retention times in RP HPLC, revealing reversible hydrate formation [7]. Most of these drugs contain polar functional groups, including amides or sulfonamides. A literature search showed that the formation of hydrates both in the solid state and in aqueous solutions is the typical chemical property of such compounds [23–29]. Therefore, we have specially synthesized a series of monofunctional N-substituted *p*-toluenesulfonamides [CH_3-C_6H_4-SO_2-NRR′ (R, R′ = H, -CH_2CH=CH_2; (I); -$(C_2H_5)_2$ (II); H, *tert*-C_4H_9 (III); H, -C_6H_5 (IV); H, -$CH_2C_6H_5$ (V); H, -C_6H_{13} (VI)] as appropriate model objects that can form hydrates in an eluent.

A comparison of the structures of gefitinib (I), pazopanib (II), and imatinib (III) (Figure 1) shows that the structure (I) contains no sulfonamide or amide groups, and it exhibits no anomalies in the recurrent approximation of the retention times (Figure 1a). Structure (II) demonstrates the deviations from linearity in the right part of the plot, corresponding to long retention times for the eluents with high water content (Figure 1b). However, structure (III), on the contrary, exhibits a recurrent anomaly in the area of the small retention times corresponding to the low water content of the eluent. This example deserves special comment because this anomaly is obviously not related to hydration.

Imatinib is a complex polyfunctional compound. Its molecule contains at least four possible nonconjugated sites for proton location. Hence, it is characterized by at least four pK_a values. They are (both experimental and precalculated (ChemAxon) values) 8.1–8.3, 3.7–4.0, 2.5, and 1.5. For our further consideration, it is important that some of these precalculated values may differ from the experimental data by 0.5–1.0.

Thus, we have a compound with one pK_a value of approximately 2.5, which is close to the pH of the eluent, 2.7–2.9. This means that two forms of this analyte (nonprotonated and protonated) exist in equilibrium in the solution:

$$X + H^+ \rightleftarrows XH^+ \tag{13}$$

Moreover, increasing the concentration of the organic component of a solvent usually leads to an increase in the pK_a values of the dissolved compounds, which can be illustrated by the dependence of the pK_a values of 2-hydroxy-4-methyl-1,3,2-dioxaphospholane 2-oxide (trivial name propylene hydrogen phosphate) on ethanol concentration in aqueous solutions [30]:

$C(C_2H_5OH)$	0	50	80	95
pK_a	1.75	2.85	3.21	4.54

Apparently, this specific effect is responsible for the deviations in the recurrent approximations from linearity, as is the case for Imatinib (III). With an increase in the acetonitrile content of the eluent, the pK_a of this compound (about 2.5 in water solutions) moves closer to the pH $\approx$ 2.7–2.9 of the eluent. The coexistence of two forms of the analyte makes the recurrent approximation of its retention time nonlinear (see anomalies in Figure 1c).

Important information on the applicability of recurrent approximation is provided by comparing the sets of retention times of the same compounds on HPLC columns of different polarities. Figure 4a (five points) presents the retention times (45–70% *v*/*v* CH_3CN content) of N-hexyl-*p*-toluenesulfonamide measured with a nonpolar EC-C18 column using acetonitrile–water eluents. Four points (excluding the right point) fall within a straight line according to the correlation coefficient R = 0.9999. The plot in Figure 4b presents the retention times (50–85% *v*/*v* CH_3CN content) of the same sulfonamide measured with a slightly more polar EC-CN column. The six points, without the right point, correspond to a straight line with R = 0.9999. In both cases, the right points visibly deviate from the linearity. Hence, the effect observed has no relation to the column polarity and is determined only by the variations in eluent composition.

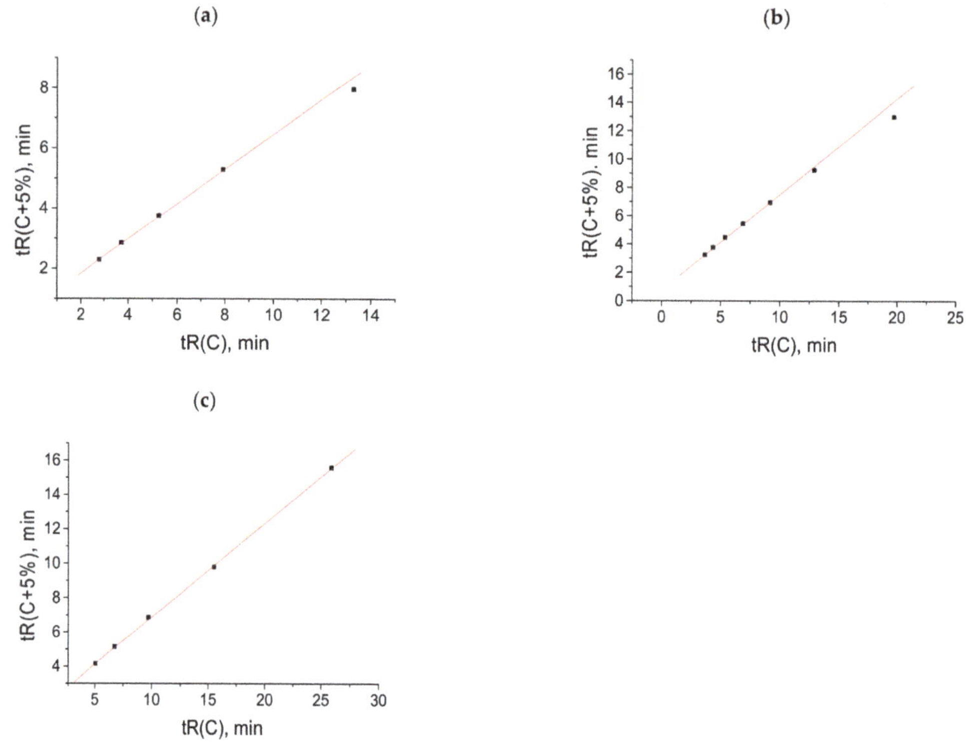

Figure 4. Recurrent approximation of the net retention times of N-hexyl-*p*-toluenesulfonamide, measured with (**a**) a 120 EC C18 column (eluent acetonitrile–water), (**b**) a 120 EC-CN column (the same eluent), and (**c**) a Phenomenex C18 column (eluent methanol–water).

The plot in Figure 4c demonstrates the recurrent approximation of the retention times measured with a nonpolar C18 column using the methanol–water eluents (55–85% v/v CH_3OH). In contrast to Figure 4a,b, the deviation of the right point from linearity is negligible (correlation coefficient for five points without the right point is 0.9998 and is $R = 0.9995$ with this point). The lower "sensitivity" of the methanol–water eluents to the formation of hydrates was discussed in [4]. Methanol itself forms rather stable monohydrates (the free energy of methanol hydration was estimated experimentally as (−5.1) kcal mol^{-1} [31]), which can effectively prevent the formation of hydrates in other compounds.

3.4. Retention Indices in Reversed-Phase HPLC: An Alternative Way to Suppose the Formation of Hydrates

The concept of retention indices (RI, Equation (4)) in reversed-phase HPLC appeared to be somewhat less popular than in gas chromatography, despite various applications [15,32,33]. This is caused by a dependence on a larger number of parameters (than in GC) (first, by the influence of different additives in the eluents) and, in general, by the narrower ranges in variation. Another reason is that the dependencies of the retention parameters of the analytes on the concentrations of organic solvents in eluents are complex [6].

The RI values for selected compounds are presented in Tables 2 and 3. Apparently, these data provide no information on the reversible formation of hydrates in an eluent. Hence, RI values should be transformed into more informative parameters.

One of the important properties of the GC retention indices is their temperature dependence, RI = $f(T)$. Usually, the presentation of this function is limited to the first term of its expansion in a Taylor series, $\beta = d\text{RI}/dT$ [15]:

$$\text{RI}(T) = \text{RI}(T_0) + d\text{RI}/dT\ (T - T_0), \quad (14)$$

where T_0 is any temperature conventionally chosen as a standard for data presentation (usually 0 or 100 °C).

This secondary parameter based on the GC retention indices depends on the differences in the topological characteristics of analytes and reference *n*-alkanes. For the majority of organic compounds, the coefficients, β, obey inequality $\beta > 0$. The $d\text{RI}/dT$ values increase with an increasing number of branches for the molecular carbon skeleton, as well as with the number and size of the rings. Specifically, the large absolute $d\text{RI}/dT$ values are responsible for the low interlaboratory reproducibility of GC retention indices [34].

The analog of the temperature dependence of the GC retention indices in RP HPLC is the dependence of the indices on the concentration of the organic solvent in an eluent, $d\text{RI}/dC$ (Equation (15)). Unlike gas chromatography, the coefficients $d\text{RI}/dC$ can be either greater or less than zero. Two dependencies RI = $f(C)$ are plotted in Figure 5 for toluene (a, $d\text{RI}/dC > 0$) and for N-phenyl-*p*-toluenesulfonamide (b, $d\text{RI}/dC < 0$). In both cases, good linearity is observed (a, $R = 0.9998$; b, $R = -0.998$); the deviations from linearity for some analytes are caused by their tautomeric transformations or prototropic equilibria.

$$\text{RI}(C) = \text{RI}(C_0) + d\text{RI}/dC\ (C - C_0), \quad (15)$$

where C is any concentration of an organic modifier chosen as a standard for data presentation (conventionally $C_0 = 0$).

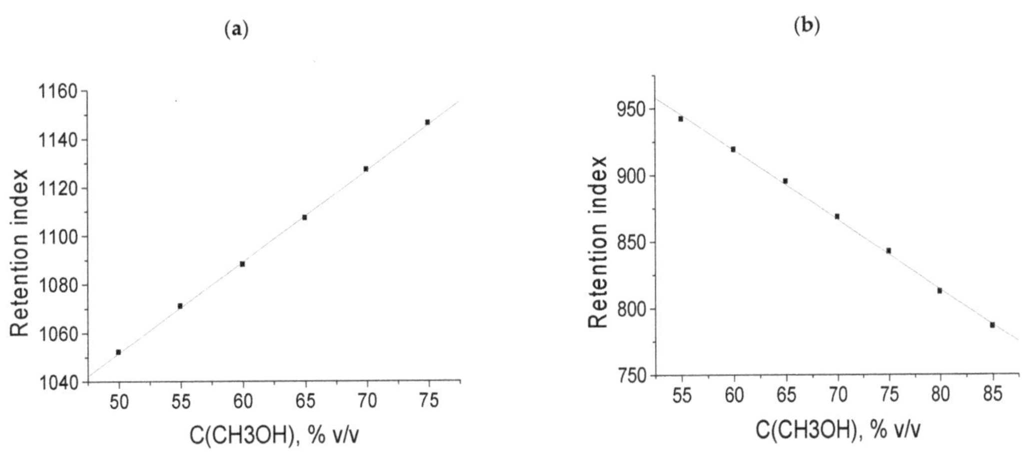

Figure 5. Two examples of the dependencies of the retention indices in RP HPLC on the concentration of methanol in an eluent: (**a**) toluene and (**b**) N-phenyl-*p*-toluenesulfonamide.

It should be noted that not the RI values and, specifically, the coefficients $d\text{RI}/dC$ can be considered for additionally confirming the formation of the hydrates of analytes in an eluent. Comparing these coefficients (data are presented in Table 6) shows that the minimal values of $d\text{RI}/dC$ belong to the most polar analytes, such as the N-substituted *p*-toluenesulfonamides, and the maximal values belong to less polar analytes, such as hydrocarbons (toluene, *o*-xylene) and their chloroderivatives (chlorobenzene). Table 6 presents the $d\text{RI}/dC$ data for selected compounds, listed in increasing order and subdivided into three subgroups: low ($d\text{RI}/dC \leq -1.0$), close to zero ($-0.4 \leq d\text{RI}/dC \leq 0.3$), and high

(≥1.6). The first subgroup (nine most polar compounds) constitutes six sulfonamides with polar fragments $–SO_2–N<$, one amide (–CO–N<), one cyclic hydrazide (–CO–NH–N<), and nitrophenol. The third subgroup includes only nonpolar compounds. Thus, we can conclude that the main factor that determines the sign and absolute values of the coefficients dRI/dC is the polarity of the analytes. The most negative values belong to the most polar sulfonamides, for which the probability of hydrate formation is maximal.

Table 6. Coefficients of the dependencies of the retention indices vs. the concentrations of the organic components in the eluents. All analytes are listed in the order of increasing $dRI/dC(CH_3OH)$ values.

Analyte	MW	*N*{H}	*d*RI/*dC*(CH_3OH)	*d*RI/*dC*(CH_3CN)	log*P* *
		Compounds with $dRI/dC << 0$			
N-Hexyl-*p*-toluenesulfonamide	255	1	−5.6 ± 0.3	−3.1 ± 0.2	4.09 ± 0.30
N-Benzyl-*p*-toluenesulfonamide	261	1	−5.3 ± 0.2	−3.8 ± 0.1	3.21 ± 0.32
N-Phenyl-*p*-toluenesulfonamide	247	1	−5.2 ± 0.1	−4.5 ± 0.2	3.04 ± 0.29
N-*tert*-Butyl-*p*-toluenesulfonamide	227	1	−4.1 ± 0.2	−2.0 ± 0.1	2.66 ± 0.32
1-Phenylpyrazolidin-3-one	162	1	−4.0 ± 0.2	−2.5 ± 0.5	0.89
N-Allyl-*p*-toluenesulfonamide	211	1	−3.4 ± 0.1	−2.0 ± 0.1	2.26 ± 0.32
N,N-Diethyl-*p*-toluenesulfonamide	227	0	−3.2 ± 0.1	−1.8 ± 0.1	2.87 ± 0.28
Diethyl-*m*-toluamide	191	0	−2.0 ± 0.1	−1.2 ± 0.3	2.18
3-Nitrophenol	139	1	−1.0 ± 0.1	−1.9 ± 0.3	2.00
		Compounds with $dRI/dC \approx 0$			
Sulphamethoxazol (hydrate)	253	3	−0.4 ± 0.1	−1.4 ± 0.3	0.89
1H-Benzortriazole (probably, hydrate)	119	1	−0.24 ± 0.04	−2.2 ± 0.4	1.44
Phthalimide (probably, hydrate)	147	1	−0.24 ± 0.04	−1.5 ± 0.3	1.15
Acetophenone hydrazone	134	2	−0.11 ± 0.05	-	1.28 ± 0.51
Acetophenone	120	0	0.0	0.0	1.70
Ninhydrin (hydrate)	178	2	0.05 ± 0.08	-	0.67
Nitrobenzene	123	0	0.3 ± 0.2	0.5 ± 0.2	1.83
2,3,5-Trimethylphenol	136	1	0.3 ± 0.1	−1.5 ± 0.2	2.73
		Compounds with $dRI/dC >> 0$			
Chlorobenzene	112	0	2.8 ± 0.2	0.8 ± 0.4	2.90
Toluene	92	0	4.0 ± 0.1	0.6 ± 0.2	2.71
o-Xylene	106	0	4.6 ± 0.3	0.2 ± 0.3	3.12

(*) Precalculated log*P* values are indicated with standard deviations (ACD software).

The easy hydration of sulfonamides and, to a lesser extent, of amides can be explained by the formation of two hydrogen bonds in the ring. This ring contains a S=O double bond (two π-electrons) and two pairs of *p*-electrons located at the N- and O-atoms (in total, sixπ- and *p*-electrons). In accordance with ($4n + 2$) Huckel's rule, such systems exhibit pseudo-aromatic properties:

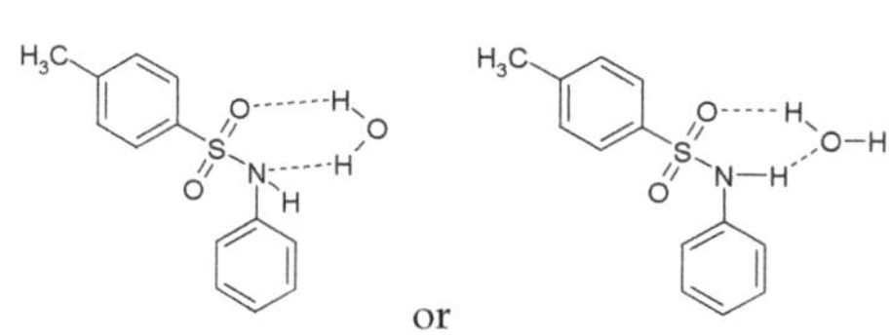

The set of compounds in the middle subgroup seems to be rather unusual. It contains four medium-polarity analytes (acetophenone, acetophenone hydrazone, nitrobenzene, and trimethylphenol) and four polar compounds: sulfamethoxazole (stable hydrate exists), ninhydrin (the same), 1H-benzotriazole, and phthalimide (formation of hydrates is rather probable). At the same time, the absolute values of the coefficients dRI/dC are not as large as those for the analytes of the first subgroup. It is interesting to note that, for acetonitrile-containing eluents, the dRI/dC values for sulfamethoxazole, 1H-benzotriazole, phthalimide, and trimethylphenol are less than -1.4, which corresponds to the compounds that are able to form hydrates. If the main reason for large negative dRI/dC values is the strong dependence of the equilibrium (of hydration equation (1)) on the content of the organic solvent in the eluent, then a lack of such dependence may be caused by the fact that the position of this equilibrium is independent of the solvent composition. In other words, the hydrate forms of some analytes from this subgroup exist under different compositions of the solvent. For example, ninhydrin (the parent structure contains no active hydrogen atoms) forms such a stable hydrate that it was characterized by pK_a 8.47, like typical organic acids [7]:

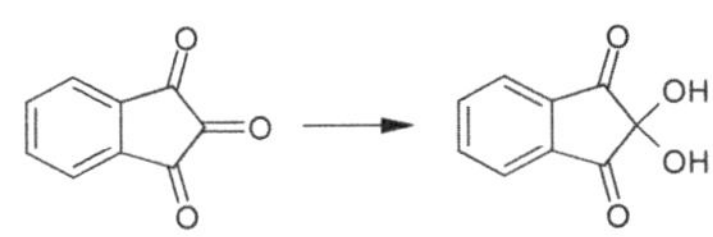

In order to finalize the consideration of the dependencies of the retention indices on the concentration of the organic constituents of an eluent, the following should be noted:

- The dRI/dC coefficients for the same compounds are not equal to each other in methanol- and acetonitrile-containing eluents. Nevertheless, for the entire set of compounds, their values satisfactorily correlate with each other (correlation coefficient R is approximately 0.87);
- The values for dRI/dC depend on the polarity of the organic compounds but show no correlation with hydrophobicity factors (log$P$), the number of active hydrogen atoms in a molecule, or the retention indices (RI);
- For compounds of a different chemical nature, the values for dRI/dC are usually different. This means that if we need to improve the separation of two analytes in a different homologous series, we can slightly change the ratio of the organic and water components of the eluent. However, this recommendation may be ineffective if such a problem arises for compounds that are similar in nature (isomers or homologs).

4. Materials and Methods

4.1. Analytes, Reagents, and Solvents

The following compounds were used: toluene, *p*-xylene, chlorobenzene, nitrobenzene (all of reagent grade, for chromatography, Reakhim, Moscow, Russia), 1-phenylpyrazolidin-3-one (reagent grade, Reakhim, Moscow, Russia), 1H-benzotriazole (for photography, Reanal, Budapest, Hungary), acetophenone, propiophenone, butyrophenone (Sigma–Aldrich Rus, LLC, Russia), 2,3,5-trimethylphenol [Theodor Schuchardt, Munich, Germany (the sample from plant volatile compounds collection of Ph.D. S. Kozhin, Leningrad State University], 3-nitrophenol (indicator, British Drug Houses, Ltd., Great Britain), and *m*-toluilic acid diethylamide (DETA, insect repellent, TU (Technical Specification) 2386-077-00205357-2007). All the selected analytes were chosen so that their pK_a values did not coincide with the pH of eluents. Some synthetic antitumor drugs discussed in the text [gefitinib (I), pazopanib (II), and imatinib (III)] were produced by BIOCAD JSC (St. Petersburg, Russia) and are characterized in [7].

The series of N-alkylsubstituted *p*-toluenesulfonamides was synthesized by Ph.D. Tatiana A. Kornilova (St. Petersburg State University) from the corresponding amines and *p*-toluenesulfonyl chloride [35].

$$CH_3\text{-}C_6H_4\text{-}SO_2Cl + 2HNRR' \rightarrow CH_3\text{-}C_6H_4\text{-}SO_2\text{-}NRR' + RR'NH\cdot HCl$$

R, R′ = H, $-CH_2CH=CH_2$; (I); $(C_2H_5)_2$ (II); H, *tert*-C_4H_9 (III); H, $-C_6H_5$ (IV); H, $-CH_2C_6H_5$ (V); H, $-C_6H_{13}$ (VI).

The reaction mixtures were analyzed directly because excess amounts of amines and their salts do not hinder the UV detection of reaction products, which (except aniline) do not absorb in the near-UV region. The presence of certain amounts of *p*-toluenesulfonic acid (in the form of the anion) follows from the appearance of peaks in the region of the retention time of the non-sorbable component.

The stock solutions of all the analytes or reaction mixtures were prepared in 2-propanol (reagent grade, Kriokhrom, St. Petersburg, Russia) and were additionally diluted with an eluent for HPLC. To prepare eluents, we used deionized water (resistivity 18.2 MΩ cm) prepared using a Milli-Q device (Millipore, USA), acetonitrile (99.5%, HPLC-gradient grade, PanReac, Spain), and methanol (analytical grade, Kriokhrom, St. Petersburg, Russia). Some eluents contained 0.1% formic acid (98% analytical grade, PanReac, Spain) or 0.1% trifluoroacetic acid. Acetonitrile-containing eluents were degassed via filtration under vacuum and sonication in a 420 W Sapfir TTTs unit (Sapfir, Russia).

4.2. Conditions of HPLC Analysis

Chromatographic analyses of both individual analytes and reaction mixtures were performed in three regimes:

(A): Agilent 1260 Infinity liquid chromatograph with a diode-array detector (scanning range 220–340 nm) and an Infinity Lab Poroshell 120 EC-C18 column 50 mm long and 3.0 mm in diameter with a sorbent particle size of 2.7 μm in water–acetonitrile mobile phases in several isocratic modes with 5% concentration steps of the organic component at an eluent flow rate of 0.4 mL min^{-1} and a column temperature of 40 °C. For the analyses of the drugs, trifluoroacetic acid was added to the eluent to a 0.1% concentration; the pH of the eluent with 50% acetonitrile content was 2.7–2.9. All the model compounds were analyzed without any acidic or salt additives added to the eluent (pH of eluents about 5.6). Samples were injected using an SN G1329A autosampler; the sample volume was 5 μL.

(B): The same chromatograph (at the same scanning range) with Agilent Poroshell 120 EC-CN columns 100 mm long and 3.0 mm in diameter with a sorbent particle size of 2.7 μm in water–acetonitrile mobile phases in several isocratic modes with 5% concentration steps of the organic component at an eluent flow rate of 0.5 mL min^{-1} and column temperature of 40 °C. The samples were injected using an SN G1329A autosampler; the sample volume was 5 μL.

(C): Shimadzu LC-20 Prominence liquid chromatograph with a diode-array detector (scanning range 190–800 nm) and Phenomenex C18 columns 250 mm long and 4.6 mm i.d. with a sorbent particle size of 5 μm in water–methanol mobile phases with the addition of 0.1% formic acid (pH of aqueous solution was 5.6) in several isocratic modes with 5% concentration steps of the organic component at an eluent flow rate of 1.0 mL min^{-1} and column temperature of 30 °C. The samples were injected using a SIL-20A/AC autosampler; the sample volume was 20 μL.

All the samples for the analyses were prepared by dissolving individual compounds or reaction mixtures in the mobile phase. The number of replicate injections of each sample in all the regimes (A)–(C) was 2–3. The interinjection variations of the retention times of the target analytes in all the cases did not exceed 0.01–0.02 min. To determine the retention indices, a mixture of three reference *n*-alkyl phenyl ketones $C_6H_5COC_nH_{2n+1}$ with n = 1–3 was added to all the samples.

4.3. Data Processing

Chromatograms in regimes (A) and (B) were obtained, processed, and stored using the Mass Hunter software (Agilent Technologies, USA). The data were statistically processed using Excel software (Microsoft Office, 2010). Origin software (versions 4.1 and 8.1) was used for calculating the parameters of the recurrent dependencies and plotting all the dependencies. The logarithmic retention indices in the isocratic regimes were calculated using Excel software or manually (with calculators).

5. Conclusions

For some polar organic compounds, we can guess the reversible formation of their hydrates during reversed-phase HPLC separation, $X + H_2O \rightleftarrows X \times H_2O$. However, hydration confirmation seems to be a complex problem. The testing of the so-called relative optical densities, $A_{rel} = A(\lambda_1)/A(\lambda_2)$, shows their dependence on the composition of eluents in some cases, but, in general, they exhibit inapplicability to the detection of hydrate formation.

One of the methods to detect the formation of hydrates seems to be the recurrent approximation of the net retention times of analytes, $t_R(C + \Delta C) = at_R(C) + b$, where ΔC is the constant step in the variations of the organic modifier content of an eluent. In the case of hydrate formation, such dependencies deviate from linearity for large retention times, e.g., for eluents with high water content.

The coefficients that characterize the dependence of the retention indices on the concentration of the organic component in an eluent, $d\mathrm{RI}/dC$, are suggested to represent an additional criterion for revealing the hydration of analytes during their reverse-phase HPLC analysis. The values of these coefficients for nonpolar compounds are largely positive ($d\mathrm{RI}/dC > 0$), whereas, for most polar analytes, they are largely negative ($d\mathrm{RI}/dC < 0$). The compounds of the latter type can, by themselves, form hydrates in HPLC eluents.

Is it possible to correct the anomalies caused by the formation of hydrates? The simplest possible way to not exclude these anomalies and minimize them is to replace acetonitrile in an eluent with methanol.

Author Contributions: D.A.N. provided all HPLC experiments with acetonitrile–water eluents; A.D. conducted all the experiments with methanol–water eluents; I.G.Z. is the supervisor of this work. All authors have read and agreed to the published version of the manuscript.

Funding: This research received no external funding.

Data Availability Statement: All data will be included into PhD Theses of A. Derouiche and D. Nikitina and will be available after finalizing these works.

Acknowledgments: All HPLC experiments with methanol as the eluent component were performed at the Center of Chemical and Material Research, Research Park of St. Petersburg State University. The authors are grateful to the staff of this Center for their assistance.

Conflicts of Interest: The authors declare no conflict of interest.

Sample Availability: Samples of the compounds are available from the authors.

References

1. Snyder, L.R.; Kirkland, J.J.; Dolan, J.W. *Introduction to Modern Liquid Chromatography*, 3rd ed.; Wiley & Sons, Inc.: New York, NY, USA, 2009; 912p.
2. Lide, D.R. (Ed.) *Handbook of Chemistry and Physics*; Taylor & Francis: Boca Raton, FL, USA, 2006.
3. Zenkevich, I.G.; Nikitina, D.A.; Derouiche, A. Formation and chromatographic detection of organic compounds hydrates. *J. Analyt. Chem. Rus.* **2021**, *76*, 493–502. [CrossRef]
4. Zenkevich, I.G.; Derouiche, A.; Nikitina, D.A. Detection of organic hydrates in reversed phase high performance liquid chromatography using recurrent approximation of their retention. *J. Liquid Chromatogr. Relat. Technol.* **2021**, *44*, 588–598. [CrossRef]
5. Lanin, S.N.; Nikitin, Y.S. The model of retention in HPLC with binary mobile phase. *Pure Appl. Chem.* **1993**, *65*, 2281–2286. [CrossRef]
6. Komsta, L. A new general equation for retention modeling from the organic modifier content of the mobile phase. *Acta Chromatogr.* **2010**, *22*, 267–279. [CrossRef]
7. Zenkevich, I.G.; Nikitina, D.A. Features of the recurrent approximation of retention parameters of polyfunctional compounds in reversed-phase high performance liquid chromatography. *Rus. J. Phys. Chem. A* **2021**, *95*, 395–402. [CrossRef]
8. Zapala, W. Influence of mobile phase composition on retention factors in different HPLC systems with chemically bonded stationary phases. *J. Chromatogr. Sci.* **2003**, *41*, 289–294. [CrossRef]

9. Snyder, L.R.; Dolan, J.W. Optimizing selectivity during reversed-phase high performance liquid chromatography method development: Prioritizing experimental conditions. *J. Chromatogr. A* **2013**, *1302*, 45–54. [CrossRef]
10. Saifutdinov, B.R.; Davankov, V.A.; Il'in, M.M. Thermodynamics of the sorption of 1,3,4-oxadiazole and 1,2,4,5-tetrazine derivatives from solutions on hypercrosslinked polystyrene. *J. Phys. Chem. A* **2014**, *88*, 358–364. [CrossRef]
11. Saifutdinov, B.R.; Buryak, A.K. The influence of water-acetonitrile solvent composition on thermodynamic characterization of adsorption of aromatic heterocycles on octadecyl-bonded silica gels. *Colloid J.* **2019**, *87*, 754–761. [CrossRef]
12. Macko, T.; Berek, D. Pressure effects in HPLC: Influence of pressure and pressure changes on peak shape, base line, and retention volume in HPLC separations. *J. Liquid Chromatogr. Relat. Technol.* **2001**, *24*, 1275–1293. [CrossRef]
13. Podgornik, A. Pressure drop in liquid chromatography. *J. Sep. Sci.* **2019**, *42*, 72–88. [CrossRef] [PubMed]
14. Zenkevich, I.G.; Derouiche, A.; Nikitina, D.A.; Kornilova, T.A.; Khakulova, A.A. Controlling the correctness of retention parameters variations in reversed phase HPLC using recurrent relations. *Anal. Control* **2021**, *25*, 117–125. (In Russian) [CrossRef]
15. Cazes, J. (Ed.) Kovats' Retention Index System. In *Encyclopedia of Chromatography*, 3rd ed.; CRC Press (Taylor & Francis Group): Boca Raton, FL, USA, 2010; Volume 2, pp. 1304–1310.
16. Peterson, M.L.; Hirsch, J. A calculation for locating the carrier gas front of a gas-liquid chromatogram. *J. Lipid Res.* **1959**, *1*, 132–134. [CrossRef]
17. Zenkevich, I.G.; Kosman, V.M. Relative absorption at various wavelengths—Additional UV-spectral parameter for identification of organic compounds in RP HPLC. *J. Anal. Chem. Rus.* **1996**, *51*, 870–875.
18. Zenkevich, I.G.; Kosman, V.M. New possibilities of identification of organic compounds by their UV-spectra using relative optical densities. *J. Appl. Chem. Rus.* **1997**, *70*, 1861–1869. (In Russian)
19. Baram, G.I.; Reikhart, D.V.; Goldberg, E.D.; Isotov, B.N.; Rodinko, M.O.; Khazanov, V.A. New possibilities of high performance liquid chromatography in pharmacopoeia analysis. *Bull. Exp. Biol. Med.* **2003**, *135*, 75–79. (In Russian) [CrossRef]
20. Zenkevich, I.G. Application of Recurrent Relations in Chemistry. *J. Chemometr.* **2010**, *24*, 158–167. [CrossRef]
21. Zenkevich, I.G. Application of Recurrent Relationships in Chromatography. *J. Chemometr.* **2009**, *23*, 179–187. [CrossRef]
22. Row, K.H. Comparison of retention models for the dependence of retention factors on mobile phase composition in reversed phase high-performance liquid chromatography. *J. Chromatogr. A* **1998**, *797*, 23–31. [CrossRef]
23. Kobayashi, M.; Nishioka, K. Hydration of N-monosubstituted amides in the binary solvents dioxane-D_2O and dioxane-H_2O. *J. Phys. Chem.* **1987**, *91*, 1247–1251. [CrossRef]
24. Guo, M.-L. 4-Acetamido-N-(4-methylpiperazinyl)benzenesulfonamide monohydrate. *Crystallogr. Commun.* **2004**, *60*, 574–575. [CrossRef]
25. Suchetan, S.F.; Foro, S.; Gowda, B.T.; Prakash, M.S. 4-Chloro-N-(3-methylbenzoyl)benzenesulfonamide monohydrate. *Acta Crystallogr. Sect. E Struct. Rep. Online* **2012**, *E68*, o46. [CrossRef]
26. Kompella, A.; Kasa, S.; Balina, V.S.; Kusumba, S.; Adibhatla, B.R.K.; Muddasani, P.R. Large scale process to prepare highly pure Bosentan monohydrate. *Sci. J. Chem.* **2014**, *2*, 9–15. [CrossRef]
27. Delgado, D.R.; Pena, M.A.; Martinez, F. Preferential solvation of some sulfonamides in 1,4-dioxane + water co-solvent mixtures at 298.15 K according to the inverse Kirkwood-Buff integrals method. *Rev. Acad. Colomb. Cienc.* **2014**, *38*, 104–114.
28. Jatezak, M.; Sidoryk, K.; Kossykowska, M.; Luniewski, W.; Zagrodzka, J.; Lipiec-Abramska, E. Development and validation of a UHPLC UV method for the in-process control of bosentan monohydrate synthesis. *Chromatographia* **2016**, *79*, 1131–1141. [CrossRef]
29. Tan, S.L.; Tiekink, E.R.T. N,N'-Bis(pyridine-3-ylmethyl)ethanediamide monohydrate: Crystal structure, Hirshfield surface analysis and computational study. *Acta Cryst.* **2020**, *E76*, 25–31. [CrossRef]
30. Edmundsen, R. (Ed.) *Dictionary of Organophosphorous Compounds*; CRC Press: London, UK, 1987.
31. Zhong, Y.; Warren, G.L.; Patel, S. Thermodynamic and structural properties of methanol-water solutions using nonadditive interaction models. *J. Comput. Chem.* **2008**, *29*, 1142–1152. [CrossRef] [PubMed]
32. Zenkevich, I.G.; Kochetova, M.V.; Larionov, O.G.; Revina, A.A.; Kosman, V.M. Retention indices as the most reproducible retention parameters in reversed phase HPLC. *Calculation for hydrophilic phenolic compounds using reference n-alkyl phenyl ketones. J. Liquid Chromatogr. Relat. Technol.* **2005**, *28*, 2141–2162. [CrossRef]
33. Smith, R.M. Retention indices in reversed-phase HPLC. In *Advanced in Chromatography*; Giddings, J.C., Grushka, E., Eds.; CRC Press: Boca-Raton, FL, USA, 2021; pp. 277–319. [CrossRef]
34. The NIST Mass Spectral Library (NIST/EPA/NIH EI MS Library, 2020 Release). Software/Data Version; NIST Standard Reference Database, Number 69, May 2020. National Institute of Standards and Technology: Gaithersburg, MD, USA, 2020. Available online: http://webbook.nist.gov (accessed on 10 December 2022).
35. Kornilova, T.A.; Derouiche, A.; Zenkevich, I.G. Recurrent approximation of retention parameters as a method for confirmation of the formation of hydrates of N-sub stituted sulfonamides in reversed phase HPLC. *Anal. Control* **2020**, *24*, 315–322. (In Russian) [CrossRef]

molecules

Article

Rapid Simultaneous Quantification of 1-Formyl-2,2-Dimethylhydrazine and Dimethylurea Isomers in Environmental Samples by Supercritical Fluid Chromatography–Tandem Mass Spectrometry

Denis V. Ovchinnikov [1], Sergey A. Vakhrameev [1], Danil I. Falev [1], Nikolay V. Ul'yanovskii [1,2,*] and Dmitry S. Kosyakov [1,*]

1 Laboratory of Environmental Analytical Chemistry, Core Facility Center "Arktika", Northern (Arctic) Federal University, Arkhangelsk 163002, Russia
2 Federal Center for Integrated Arctic Research, Arkhangelsk 163000, Russia
* Correspondence: n.ulyanovsky@narfu.ru (N.V.U.); d.kosyakov@narfu.ru (D.S.K.)

Abstract: When released to the environment, the rocket fuel unsymmetrical dimethylhydrazine (UDMH) undergoes oxidative transformations, resulting in the formation of an extremely large number of nitrogen-containing transformation products, including isomeric compounds which are difficult to discriminate by common chromatography techniques. In the present work, supercritical fluid chromatography–tandem mass spectrometry (SFC-MS/MS) was proposed for resolving the problem of fast separation and simultaneous quantification of 1-formyl-2,2-dimethylhydrazine (FADMH) as one of the major UDMH transformation products, and its isomers—1,1-dimethylurea (UDMU) and 1,2-dimethylurea (SDMU). 2-Ethylpyridine stationary phase provided baseline separation of analytes in 1.5 min without the distortion of the chromatographic peaks. Optimization of SFC separation and MS/MS detection conditions allowed for the development of rapid, sensitive, and "green" method for the simultaneous determination of FADMH, UDMU, and SDMU in environmental samples with LOQs of 1–10 μg L^{-1} and linear range covering three orders of magnitude. The method was validated and successfully tested on the real extracts of peaty and sandy soils polluted with rocket fuel and UDMH oxidation products. It was shown that both UDMU and SDMU are formed in noticeable amounts during UDMH oxidation. Despite relatively low toxicity, UDMU can be considered one of the major UDMH transformation products and a potential marker of soil pollution with toxic rocket fuel.

Keywords: formic acid dimethylhydrazide; dimethylurea; rocket fuel; transformation products; supercritical fluid chromatography; tandem mass spectrometry

1. Introduction

Despite the increasingly active use of environmentally friendly types of rocket propellants (kerosene, methane, and hydrogen in combination with liquid oxygen as an oxidizer), the space programs of different countries still rely on the operation of launch vehicles or booster blocks using toxic unsymmetrical dimethylhydrazine (UDMH) as a fuel [1,2]. When released to the environment, UDMH rapidly undergoes oxidative transformations via radical mechanism resulting in the formation of extremely large number (up to one thousand) of nitrogen-containing transformation products including toxic and carcinogenic compounds [3,4]. To date, several dozen UDMH transformation products have been reliably identified [5–7], the most abundant of them are *N*-nitrosodimethylamine (NDMA), formaldehyde dimethylhydrazone, 1,1,4,4-tetramethyltetrazene, *N*,*N*-dimethylformamide, dimethylaminoacetonitrile, 1-methyl-1H-1,2,4-triazole, and 1-formyl-2,2-dimethylhydrazine (formic acid *N*′,*N*′-dimethylhydrazide, FADMH) [8–13]. The latter compound has been identified as an UDMH transformation product relatively recently [9], although it is always found in significant amounts in soils and waters contaminated with rocket fuel [12,14,15]. Analyses of UDMH oxidation products in model laboratory experiments and real soil samples by atmospheric pressure ionization high-resolution mass spectrometry [4] showed the presence, in all mass spectra, of an intense signal of the compound $C_3H_8N_2O$, which can be attributed to FADMH. However, the use of hydrogen/deuterium isotopic exchange mass spectrometry allowed the discrimination of at least three structures with the indicated elemental composition and suggestion of the presence of *N,N*-dimethylurea (unsymmetrical dimethylurea, UDMU) among them [3]. This assumption was confirmed by the published data indicating the identification of UDMU by GC-MS [1]. Another but less probable product with the same elemental composition is *N,N′*-dimethylurea (symmetrical dimethylurea, SDMU), the discovery of which among UDMH transformation products has not yet been reported in the literature. Since the available information on the possibility of the formation of substantial amounts of UDMU and SDMU along with FADMH during the oxidative transformation of rocket fuel is extremely scarce and requires additional studies, the development of approaches to the simultaneous quantification of these isomeric compounds in complex objects is of great interest. The solution of this problem is complicated by the high polarity and similarity of the physicochemical properties of FADMH, UDMU, and SDMU (Table 1).

Table 1. List of analytes and their physicochemical properties.

Analyte	CAS Number	Structural Formula	Molecular Weight, Da	pKa *	Log*P*
1-formyl-2,2-dimethylhydrazine (FADMH)	3298-49-5	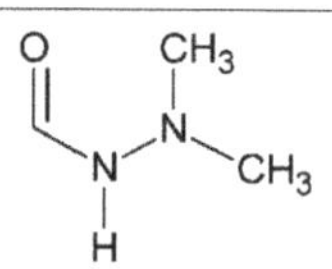	88.1	3.5 ± 0.7 **	−0.81 ± 0.53 **
N,N-dimethylurea (UDMU)	598-94-7	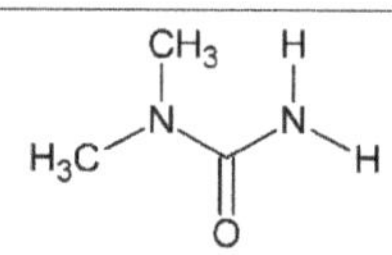	88.1	−0.2 ± 0.7 **	−1.28 ± 0.54 **
N,N′-dimethylurea (SDMU)	96-31-1	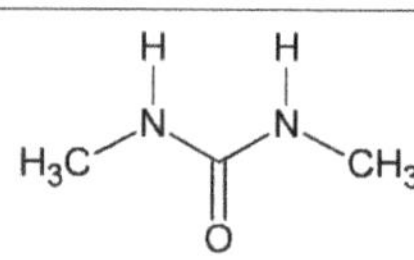	88.1	−0.6 ± 0.7 **	−1.02 ± 0.30 **

* For the protonated form. ** The values predicted in silico by ACD/Labs Percepta platform software [16].

In the case of analytes with reactive amino groups, high-performance liquid chromatography (HPLC) is considered a preferred separation technique since it does not require the tedious procedures of preliminary derivatization and matrix change. In combination with mass spectrometry (MS), it has been successfully used for quantification of hydrazines and most of the above-mentioned major UDMH transformation products [17] and made it possible to achieve the limits of FADMH quantification (LOQs) at a level of 0.01–6 μg L^{-1} [11,12]. However, in the case of polar non-ionogenic compounds as FADMH and dimethylurea isomers, the retention on commonly used reversed and ion-exchange stationary phases is relatively weak and separation selectivity is insufficient. Due to the close polarities of such analytes, the use of hydrophilic interaction liquid chromatography also does not allow for the complete separation of FADMH and its isomers and effective elimination of matrix effects.

In our opinion, the most promising method for the simultaneous determination of these compounds is supercritical fluid chromatography–mass spectrometry (SFC-MS) providing separation based on the specific interactions of analytes with polar stationary phase, which is "orthogonal" [18] to a reversed phase HPLC. The low viscosity and high diffusion coefficients of the sub- or supercritical carbon dioxide, which is used as a main component of the mobile phase in SFC, ensure high separation speed and efficiency. Even though there are no published works devoted to the use of SFC or SFC-MS for the determination of UDMH transformation products, SFC has shown itself superior to HPLC in the analysis of isomers [19,20] and various polar compounds [21–23]. The combination of SFC with tandem mass spectrometry detection (SFC-MS/MS) provides high sensitivity and selectivity of analyses of complex objects and does not require any additional specific equipment [24–26]. Currently, this analytical technique is increasingly used in practice and appears to be a promising alternative to HPLC-MS/MS. The present work is aimed at the development of a rapid and sensitive SFC-MS/MS method for the simultaneous determination of FADMH, UDMU, and SDMU in water samples and soil extracts and thus obtaining new knowledge on the rocket fuel transformation processes in the environment.

2. Results and Discussion

2.1. Mass Spectra of Analytes and Mass Spectrometry Detection

Being isomeric compounds and having the same molecular weight, all analytes give the signals of protonated molecules $[M + H]^+$ at m/z 89 under the conditions of positive ion mode atmospheric pressure ionization. Despite the evidence previously noted for SFC-MS certain gain in ionization efficiency of nitrogen-containing compounds under electrospray (ESI) conditions compared to atmospheric pressure chemical ionization (APCI) [27], in our preliminary tests, both techniques demonstrated the close intensities of $[M + H]^+$ signals at high (>1 mL min^{-1}) flowrates of the mobile phase. In this situation, an APCI technique was chosen for further method development due to its less susceptibility to matrix effects. Tandem mass spectrometry in multiple reaction monitoring (MRM) mode was used to ensure the high selectivity of analysis considering relatively low retention of analytes and thus the possibility of co-elution with matrix components. The recorded tandem mass spectra (Supplementary Materials Figure S1) demonstrate the difference in the collision induced dissociation (CID) pathways of UDMU and SDMU—the first compound predominantly eliminates NH_3 from primary amino group (m/z 89 $\rightarrow$ 72), while the second one is characterized by an easy loss of methylamine (m/z 89 $\rightarrow$ 58). These ion transitions were chosen for quantification purposes. In the case of FADMH, CID results in a cleavage of *N-N* bond with the formation of protonated *N*-methylmethanimine (m/z 44) or simultaneous loss of carbonyl and methyl groups leading to the formation of methyldiazene or formaldehyde hydrazone (m/z 45) protonated molecules. The intensity ratios of the corresponding peaks in mass spectra strongly depend on the applied collision energy. As a result of automated optimization of the collision energies for both MRM transitions, the product ion with m/z 45 was chosen as a quantifier. The optimized parameters of MRM detection are summarized in Table 2.

Table 2. Detection parameters in the multiple reaction monitoring mode.

Analyte	Precursor Ion, *m/z*	Product Ion, *m/z*	Declustering Potential, V	Collision Energy, eV
FADMH	89	45 (71 *)	20	20
UDMU	89	72 (46 *)	30	20
SDMU	89	58 (44 *)	30	30

* Qualifier ion.

2.2. Screening of SFC Stationary Phases and Optimization of Separation Conditions

The key factor affecting the retention and separation of polar analytes is the nature of the stationary phase. At the first stage of the study, silica-based stationary phases with a particle size of 1.7–3 μm (see Section 3.2) differing in the chemistry of the bonded groups were screened: bridged ethylene hybrid bare silica (Acquity BEH), three octadecyl stationary phases (endcapped Titan C18, Acquity HSS C18 with no endcapping, Nucleodur ISIS with cross-linked octadecyl groups), two sorbents with embedded polar non-ionogenic groups (Nucleodur PolarTec with amide linker between silica surface and octadecyl group, Acquity HSS Cyano with cyanopropyl bonding and high silanol activity), and two stationary phases with polar ionogenic groups (Acquity BEH 2-EP with 2-ethylpyridine moiety and Nucleodur NH2-RP with aminopropyl groups). The chromatograms obtained under common SFC conditions (backpressure 130 bar, mobile phase—10% methanol in CO_2) (Figure S2) showed potential for the rapid separation of analytes and specificity of their retention on different stationary phases. First, the difference in the behavior of FADMH and dimethylurea isomers should be noted. Despite the close polarities (Log*P*) of all analytes (Table 1) and the presence of the same functional groups, the retention of FADMH is much lower compared to UDMU and SDMU. It is also worth noting that extremely strong tailing and even splitting of FADMH peaks obtained for most stationary phases occur. This can be explained by the higher ability of this compound for protonation (acidity constant of conjugated acid is 4 orders of magnitude lower when compared to other analytes) and thus the presence in mobile phase mostly as cation due to acidic conditions in carbon dioxide–methanol mixture containing significant amounts of methylcarbonic acid [28,29]. The strong interactions of both cationic and neutral forms of FADMH with silica surface leads to the above-noted peak distortions observed mostly for the sorbents with most accessible silanol groups—Acquity BEH, HSS Cyano, and HSS C18. In contrast, the similar Nucleodur ISIS octadecyl phase with a well-shielded silica polar surface provides an acceptable FADMH peak shape, while it does not ensure sufficient retention and separation of UDMU and SDMU, for which the contribution of polar interactions with silanols is crucial. This is in a good agreement with our recent observation of polar retention of pentacyclic triterpenoids on octadecyl stationary phases in SFC at low contents (<6%) of methanol in the mobile phase [30]. An interesting fact is an inversion of symmetrical/unsymmetrical dimethylurea isomers elution order on different stationary phases. Bare silica and sorbents capable of hydrogen bonding and ion exchange (BEH 2-EP, NH2-RP, and PolarTec) provide stronger retention of SDMU, while four other stationary phases (octadecyl and cyanopropyl) are characterized by reversed elution order. While a noticeable contribution of nonpolar retention can be a reason in the case of C18 sorbents, a satisfactory explanation of this phenomenon for cyanopropyl stationary phase has not been found. Among all tested stationary phases, Acquity BEH 2-EP demonstrated best peak shapes and baseline separation of all analytes, although retention factors were relatively low. Combination of hydrogen bond donor and acceptor properties with the capability of π-π interactions allows us to consider this sorbent as universal stationary phase for SFC separations of many polar analytes. Moreover, due to the superior peak shapes even without using mobile phase additives BEH 2-EP is known as the best choice in the analyses of various basic nitrogen-containing compounds [31,32]. Thus, further method development and optimization steps in our work were carried out using Acquity BEH 2-EP chromatographic column.

It has been found that the introduction of formic acid (0.1%, v/v), ammonium formate (10 mM) and water (5% v/v) as mobile phase additives (dynamic modifiers), regulating pH and the availability of silanol groups of the sorbent, did not have a significant effect on the retention of analytes, the shape of the chromatographic peaks, and the separation selectivity. Since APCI, unlike ESI, is not sensitive to the analytes protolytic equilibria in the mobile phase, the addition of formic acid did not affect the ionization efficiency and, therefore, the sensitivity of mass spectrometric detection. In this regard, neat methanol was recommended for use as a co-solvent for carbon dioxide. With an increase in the methanol content the retention times (t_R) of all analytes expectedly decreased due to the polar retention mechanism. This factor led to a simultaneous decrease in separation selectivity. At the same time, a substantial improvement in the chromatographic peak shapes and widths was observed (Supplementary Materials Table S1). The methanol content of 10% (v/v) was found to be optimal and allowed the separation of analytes with selectivity (α) and resolution (R) of >1.5 and 2.0, respectively (Table S1).

Temperature and backpressure usually do not have a significant effect on SFC separations with polar stationary phases and are often considered as secondary parameters when optimizing a chromatographic method [33]. Indeed, we noted some improvement in the chromatographic peak shapes (especially for FADMH) and a slight decrease in retention times with an increase in backpressure, which can be partially compensated by an increase in temperature. Based on these considerations, the operating backpressure of 190 bar and column temperature of 55 °C close to the maximum possible (in terms of the SFC system performance and ensuring the lifetime of the chromatographic column) values were chosen as optimal.

Summarizing the above, the following analysis conditions can be recommended: stationary phase—Acquity UPC2 BEH 2-EP; flow rate—1.30 mL min^{-1}; methanol content in the mobile phase—10% (v/v); temperature—55 °C; backpressure—190 bar. The chromatogram of the model mixture of analytes (Figure 1) obtained under the indicated conditions demonstrates the correct peak shapes and baseline separation of analytes in the isocratic elution mode with an analysis time of 1.5 min.

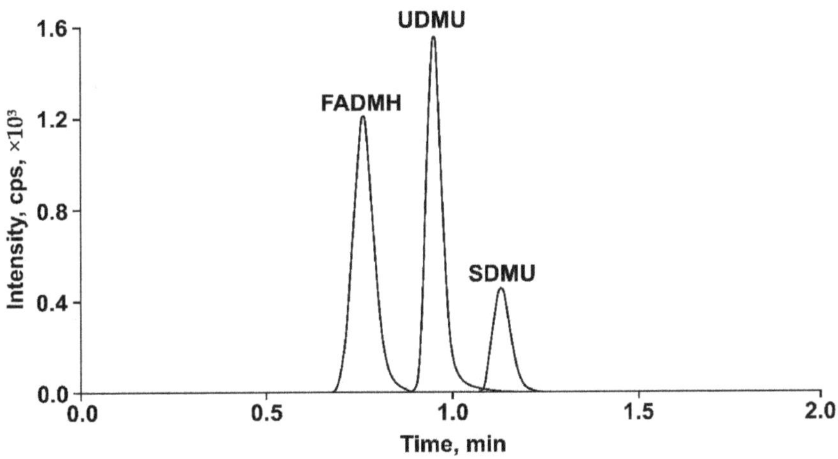

Figure 1. The SFC-MS/MS chromatogram of the model mixture of analytes (3 µg mL^{-1} of FADMH, 0.5 µg mL^{-1} of UDMU and SDMU) obtained under the optimized conditions.

2.3. Validation of the Developed Method

The attained values of instrumental limits of detection (LODs) and limits of quantification (LOQs) (Table 3) were typical for HPLC-MS/MS technique and fell within the ranges of 0.4–3 and 1.3–10 µg L^{-1}, respectively. The analyses of calibration solutions of analytes and construction of calibration dependences of peak areas (y) on concentration (x) in the form y = ax showed good linearity (R^2 > 0.999) in the concentration range covering three orders of magnitude.

Table 3. The key specifications of the developed SFC-MS/MS method.

Analyte	a	R^2	Linear Range, µg L^{-1}	LOD, µg L^{-1}	LOQ, µg L^{-1}
FADMH	170	0.9998	LOQ-6250	3.0	10
UDMU	1200	0.9995	LOQ-1000	0.4	1.3
SDMU	360	0.9995	LOQ-1000	0.5	1.7

In the case of non-aqueous samples (e.g., acetonitrile extracts of soils [34]) the LODs/LOCs can be further reduced by an increase in the sample injection volume up to 5 µL or even more which is allowed by the used chromatographic column. However, it is strongly not recommended for aqueous samples due to the observed smearing of the chromatographic peaks when injection volumes >2 μL are used. This effect is associated with competition of water and analytes for sorption centers of the stationary phase.

Inter-day and intra-day assays carried out on the model solutions at three concentration levels (from ~LOQ) within an entire linear range (Supplementary Materials Table S2) demonstrated high precision (RSD < 15%) of the developed method at the lowest concentrations of analytes. Moreover, at the level of ≥10, LOQ RSD values were below 4% even in inter-day precision test.

An accuracy of the method was evaluated by spike recovery test using real samples of river water (sample 1) and acetonitrile extract of peat bog soil (sample 2) with very complex matrix (high content of natural organic matter), not containing the studied analytes. The obtained spike recoveries (Table 4) were in the range of 90–115%, including those measured at the LOQ level. Thus, taking into account the high precision of the method, the matrix effects can be considered insignificant even in the case of peat bog water.

Table 4. Accuracy of the method determined by spike recovery test on real samples of river and beat bog water (n = 3, p = 0.95).

Analyte	Spiked, μg L^{-1}	Found, μg L^{-1}		Accuracy, %	
		Sample 1	Sample 2	Sample 1	Sample 2
FADMH	12.5	12.6 ± 1.3	14.3 ± 2.4	101 ± 11	114 ± 17
	310	314 ± 6	320 ± 24	100 ± 3	102 ± 8
	2500	2560 ± 60	2420 ± 70	102 ± 2	97 ± 3
UDMU	2.0	2.2 ± 0.3	2.3 ± 0.3	110 ± 16	115 ± 13
	50	51 ± 2	56 ± 3	102 ± 4	112 ± 5
	400	407 ± 20	410 ± 30	102 ± 5	103 ± 7
SDMU	2.0	2.0 ± 0.2	1.8 ± 0.3	100 ± 10	90 ± 17
	50	49 ± 3	48 ± 4	98 ± 6	96 ± 8
	400	390 ± 15	375 ± 40	98 ± 4	94 ± 11

The high robustness of the method is due to the use of isocratic elution ensuring the high analysis reproducibility. It was confirmed in the analyses of a great number of real samples without noticeable change in retention times or selectivity loss.

The comparison of the developed method with those described in the literature for single analytes and based on GC [14,34,35] or HPLC [12] separations showed sensitivity similar to GC-MS or HPLC-MS and at least one order of magnitude gain in LOD for SDMU when compared with the GC-NPD method provided by the U.S. Environmental Protection Agency (EPA) [36]. It is worth noting that the SFC-MS/MS method, unlike those mentioned, is distinguished with exceptional rapidity (1.5 min), low consumption of organic solvent, and low cost of the mobile phase.

2.4. Analyses of Real Samples

The developed method was successfully tested in the analyses of four real samples containing UDMH transformation products—acetonitrile extracts of soils polluted with rocket fuel and taken from the site of accidental crush of the launch vehicle near Baikonur spaceport (sample 3) and from the landing site of the launch vehicle's burned-out first stage in northern Russia (sample 4), as well as aqueous solutions of UDMH treated with ozone (sample 5) and hydrogen peroxide in the presence of Cu^{2+} ions as catalyst (sample 6). The recorded chromatograms (Figure 2) revealed the presence of all analytes with concentrations >LOQ. An exception is sample 6, in which SDMU was not detected. As expected, FADMH predominated in all samples (Table 5) and was found at the levels reaching 200 mg L^{-1} (sample 5). This value corresponds to 20% of the initial UDMH content and allows the consideration of FADMH as a main primary transformation product formed under action of ozone.

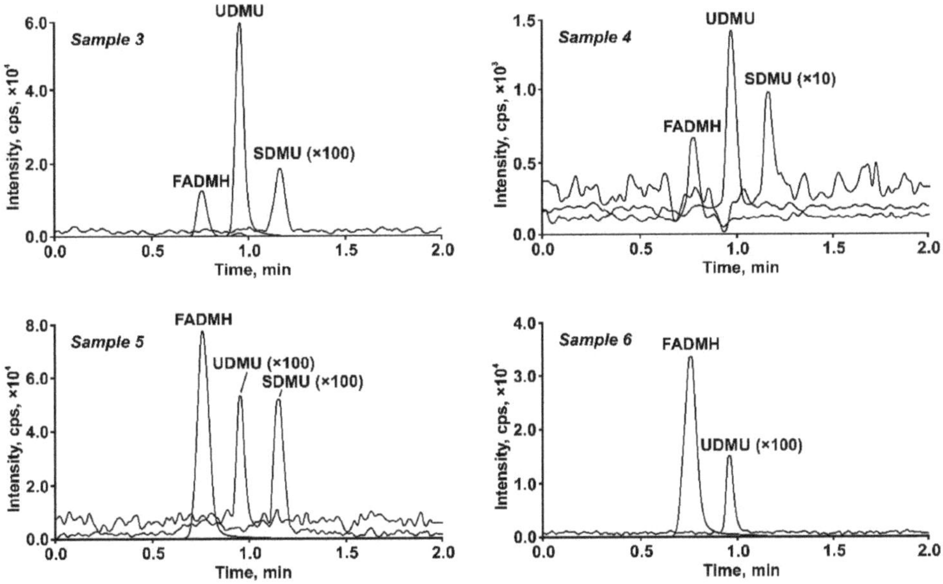

Figure 2. SFC-MS/MS chromatograms of real samples containing UDMH transformation products.

Table 5. Measured concentrations (contents) of analytes in real samples of UDMH transformation products.

Sample	FADMH	UDMU	SDMU
		Content, mg kg^{-1}	
3	190 ± 15	99 ± 3	1.2 ± 0.1
4	4.2 ± 0.6	0.95 ± 0.09	0.27 ± 0.03
		Concentration, mg L^{-1}	
5	200 ± 6	0.11 ± 0.04	0.53 ± 0.07
6	105 ± 6	0.36 ± 0.05	<LOQ

Another advanced oxidation process, based on the UDMH catalytic treatment with hydrogen peroxide, also provided the conversion of UDMH mainly to FADMH (10% of initial UDMH after 24 h reaction). The results of the model experiments on UDMH oxidation were in a good agreement with the high contents of FADMH found in the soil extracts. For example, sandy soil (sample 3) polluted with rocket fuel and subjected to reagent treatment with hydrogen peroxide contained 190 mg kg^{-1} of this transformation product. Peat bog soil sample contained lower amounts of FADMH; however, the measured content (4.2 mg kg^{-1}) was also significant considering the smaller scale of fuel spill and the known ability of peat to strongly bind UDMH and thus prevent its oxidative transformations [15].

Of great interest is the data obtained for the first time on the levels of dimethylurea isomers in polluted soils and UDMH aqueous oxidation products. In model experiments with ozone and hydrogen peroxide as oxidants (samples 5 and 6), the concentrations of UDMU and SDMU were about three orders of magnitude lower compared to FADMH, while in soil samples 3 and 4 the ratios FADMH/UDMU were only 1.9 and 4.4, respectively. In contrast, SDMU was found in less amounts in comparison with those detected in sample 5. It is worth noting that in the peat bog soil sample, UDMU and SDMU contents differed by less than four times. This means that the slow transformation of UDMH in soils provides conditions for the formation of significant amounts of UDMU, which can be considered as one of the major UDMH transformation products and an important marker of rocket fuel contamination, regardless of soil type. Another unexpected result is the formation of SDMU despite the absence of the dimethylamine group in its structure. This once again confirms the hypothesis of the radical nature of UDMH oxidative transformations, accompanied by the transfer of methyl radicals [3].

Given the prominent place of dimethylurea, along with FADMH, among UDMH transformation products, it is of considerable interest to evaluate and compare the toxicity of these isomeric compounds. The in silico prediction involving models based on quantitative structure–activity/toxicity relationships (QSAR/QSTR) revealed relatively low (Category IV) and comparable acute toxicity of all three analytes (Supplementary Materials Table S3). The similar pattern was observed for aquatic toxicity; however, it should be noted that the estimated LC_{50} values for SDMU turned out to be much lower (up to one order of magnitude) than those for FADMH and UDMU. The situation is different in the case of mutagenicity. The predicted probability of positive Ames test is quite high (0.51) for FADMH, while both dimethylurea isomers had this value below 0.20.

3. Materials and Methods

3.1. Analytes, Reagents and Materials

1,1-Dimethylurea and 1,3-dimethylurea were purchased from Alfa Aesar (Karlsruhe, Germany) and had a purity of ≥97%. 1-formyl-2,2-dimethylhydrazine was synthesized from UDMH (>98%, Sigma-Aldrich, Steinheim, Germany) and ethyl formate (97%, Sigma-Aldrich, Germany) by the known procedure [37].

In SFC-MS/MS analyses, carbon dioxide (99.99%) and HPLC gradient grade methanol purchased from Cryogen (Aramil, Russia) and Khimmed (Moscow, Russia), respectively, formic acid (>98%, Sigma-Aldrich, St. Louis, MO, USA), 10 M aqueous solution of ammonium formate (Sigma-Aldrich, St. Louis, MO, USA), and ultrapure Type I Milli-Q water were used as components of mobile phase and dynamic modifiers.

HPLC gradient grade acetonitrile (0 grade, Cryochrom, St.-Petersburg, Russia), barium hydroxide (pure, Panreac, Barcelona, Spain), sulfuric acid, and "chem. pure" grade isopropanol (Komponent-Reaktiv, Moscow, Russia) were used in the soil extraction procedure and for the preparation of analyte solutions. High purity 30–35% aqueous solutions of hydrogen peroxide and copper sulfate (>98%) purchased from Neva-Reaktiv (St.-Petersburg, Russia) were used in the procedure of UDMH oxidation.

The stock solutions of individual analytes in methanol with concentration of 2 mg mL^{-1} were prepared from accurately weighed portions and stored in a freezer at a temperature of −20 °C. Working and calibration solutions were prepared immediately before the experiments by consecutive dilutions of the mixture of stock solutions with isopropanol.

3.2. Real Objects and Sample Preparation

Two real samples, which were not contaminated with rocket fuel and thus did not contain the studied analytes, were used in the method validation procedure:

Sample 1. River water taken from the mouth of the Northern Dvina River. Salinity—140 mg L^{-1}; dissolved organic carbon content—13 mg L^{-1}.

Sample 2. Acetonitrile extract of peat bog soil, obtained by pressurized liquid extraction (PLE). The sample was taken in the northeast of the Arkhangelsk region (Russia), where most of the fall sites of launch vehicle first stages launched from the Plesetsk cosmodrome are located.

The approbation of the developed SFC-MS/MS method on real objects was carried out using two samples of soil (PLE extracts), polluted with rocket fuel, and two samples of UDMH aqueous solutions treated with oxidative reagents:

Sample 3. Acetonitrile PLE extract of sandy grey desert soil, taken from the site of the Proton launch vehicle accidental crash near Baikonur cosmodrome (Kazakhstan) in 2013 after on-site detoxification with oxidative reagent based on hydrogen peroxide and iron complexonate according to current regulations [38].

Sample 4. Acetonitrile PLE extract of peat bog soil taken at the fall site of a launch vehicle first stage in the northeast of the Arkhangelsk region (Russia). The moisture content was 89% and organic matter content was 98% (recalculated to oven-dry sample).

Sample 5. An aqueous solution of 1,1-dimethylhydrazine with an initial concentration of 1000 mg L^{-1} subjected to ozone treatment. The oxidizing agent (200 mg h^{-1}) was passed with air flow through UDMH solution (10 mL) for 20 min. Ozone was produced from dry air using Enaly 500AF portable ozone generator (Shanghai, China).

Sample 6. An aqueous solution of 1,1-dimethylhydrazine with an initial concentration of 1000 mg L^{-1} after hydrogen peroxide treatment. In total, 20 μL of 30% H_2O_2 solution was added to 10 mL of UDMH solution. To accelerate the transformation processes, an addition of copper sulfate as a catalyst was used; the concentration of copper ions in the reaction mixture was 1 mg L^{-1}.

PLE of all soil samples with acetonitrile was carried out using an ASE-350 (Dionex, Sunnyvale, CA, USA) accelerated solvent extraction system in nitrogen atmosphere according to the earlier developed procedure [33] briefly described below.

In the case of sandy soil, a thoroughly averaged sample weighing 1.0 g (dry matter) was placed into a 5 mL stainless steel extraction cell. Extraction was carried out at a pressure of 100 bar and a temperature of 100 °C in two 10 min cycles. Between cycles and at the end of the extraction, rinsing with a fresh portions of the solvent (60% of the cell volume) was used. The resulting volume of the obtained extract was ~30 mL.

In the case of peat, a sample weighing 5.0 g (0.55 g of dry matter) was thoroughly mixed with 1.25 g of barium hydroxide and placed in a 10 mL stainless steel extraction cell. Acetonitrile containing 10% of water was used as an extractant. The same extraction conditions and time program used for the sandy soil were used. The resulting volume of the obtained extract was ~50 mL. To neutralize the strongly alkaline medium and remove dissolved barium hydroxide, 1 M sulfuric acid was added to the extract until pH reached 3–5. The barium sulfate precipitate was removed by centrifugation.

Before analysis, samples 3, 5, and 6 were diluted with isopropanol 100-fold and sample 4—10-fold. After filtration through 0.22 μm nylon membrane filter, samples were injected into SFC-MS/MS system.

3.3. Supercritical Fluid Chromatography–Tandem Mass Spectrometry

Chromatographic separation was carried out on a Waters Acquity UPC^2 (Milford, MA, USA) SFC system consisted of two chromatographic pumps for carbon dioxide and co-solvent, autosampler, column thermostat, and back-pressure regulator (BPR). Make-up solvent (methanol) flow was introduced through tee connector after BPR using additional external chromatographic pump Ultimate 3000 RS (Thermo Fisher Scientific, Waltham, MA, USA). A 150 cm PEEK capillary with an internal diameter of 0.075 mm was used as an interface between the SFC system and mass spectrometer.

The following chromatographic columns were tested in stationary phase screening: Acquity UPC2 BEH, 150 × 3.0 mm, 1.7 μm (Waters, Milford, MA, USA); Acquity HSS Cyano, 150 × 3.0 mm, 1.8 μm (Waters, Milford, MA, USA); Acquity UPC2 BEH 2-EP, 150 × 3.0 mm, 1.7 μm (Waters, Milford, MA, USA); Nucleodur NH2-RP, 125 × 2.0 mm, 3.0 μm (Macherey-Nagel, Duren, Germany); Titan C18, 100 × 2.1 mm, 1.9 μm (Supelco, Bellefonte, PA, USA); Acquity UPC2 HSS C18 SB, 150 × 3.0 mm, 1.8 μm (Waters, Milford, MA, USA); Nucleodur PolarTec, 150 × 2.0 mm, 1.8 μm (Macherey-Nagel, Duren, Germany); Nucleodur C18 Isis, 150 × 2.0 mm, 1.8 μm (Macherey-Nagel, Duren, Germany).

Experiments on stationary phase screening were carried out under the following conditions: mobile phase—carbon dioxide containing 10% methanol; flow rate—1.3 mL min^{-1}; column temperature—25 °C; back pressure—130 bar; injection volume—2.0 μL; and make-up solvent flow rate—0.10 mL min^{-1}. The void volume of the chromatographic system for calculating retention factors (k) was determined from the first perturbation of the baseline.

Mass spectrometry detection was performed using an AB Sciex 3200 QTrap triple quadrupole tandem mass spectrometer (Concord, ON, Canada) equipped with Turbo-V ion source with atmospheric pressure chemical ionization (APCI) probe. The positive ion mode (APCI+) was used with further ion source parameters: corona needle current—4 μA; source temperature—300 °C; curtain; nebulizing and drying gas pressure—20; and 50 and 30 psi, respectively.

3.4. Method Validation

The LODs and LOQs of the analytes were calculated using signal-to-noise ratio (S/N) criteria of 3 and 10, respectively, and then refined in the analysis of solutions with concentrations close to LOQ. The intra-day precision (RSD) was estimated at three levels in a series of consecutive chromatographic analyses of the standard solutions (n = 10). The inter-day precision was determined in the same manner within 48 h (n = 20). The matrix effect and accuracy of analyte quantification were estimated by the spike recovery test at three concentration levels, known amounts of analytes were introduced into river water sample and peaty soil extract, followed by SFC-MS/MS analyses in three replicates.

3.5. In Silico Toxicity Prediction

Acute toxicity (mouse, rats) and toxicity towards two aquatic organisms, fathead minnow (*P. promelas*) and water flea (*D. magna*), as well as mutagenicity (probability of positive Ames test) were predicted with the ACD/Labs Percepta software v. 2021.1.3 (Advanced Chemistry Development, Inc., Toronto, ON, Canada) using quantitative structure–activity/toxicity relationship (QSAR/QSTR) algorithms.

4. Conclusions

A rapid, highly sensitive, and "green" method for the simultaneous determination of the three isomeric products (1-formyl-2,2-dimethylhydrazine, 1,3-dimethylurea, and 1,1-dimethylurea) of toxic rocket fuel transformations by supercritical fluid chromatography–tandem mass spectrometry was developed, validated, and successfully tested on real objects. The baseline separation of polar analytes was achieved on a 2-ethylpyridinium stationary phase using carbon dioxide with the addition of methanol (10%) as a mobile phase in isocratic elution mode. The attained detection limits are in the range of 0.4–3.0 μg L^{-1}, while the linear concentration range covers three orders of magnitude. The method is distinguished by exceptionally short analysis time (1.5 min), low consumption of organic solvents, and low operational cost. Its application to the study of peaty and sandy soils polluted with rocket fuel as well as complex mixtures of 1,1-dimethylhydrazine oxidation products for the first time allowed the estimation of the levels of two dimethylurea isomers and confirmation of their formation in polluted soils. Despite relatively low toxicity, 1,1-dimethylurea can be considered one of the major 1,1-dimethylhydrazine transformation products and a potential marker of soil pollution with toxic rocket fuel due to its high content in soil samples.

Supplementary Materials: The following supporting information can be downloaded at: https://www.mdpi.com/article/10.3390/molecules27155025/s1, Figure S1: Tandem mass spectra of FADMH, UDMU, and SDMU. Collision energy—30 eV with spread of 20 eV; Figure S2: SFC-MS/MS chromatograms of the model mixture of analytes obtained for different stationary phases (Acquity BEH, Titan C18, Acquity HSS C18, Nucleodur ISIS, Nucleodur PolarTec, Acquity HSS Cyano, Acquity BEH 2-EP, Nucleodur NH2-RP). Analysis conditions: mobile phase—carbon dioxide with 10% of methanol; flowrate—1.3 mL min^{-1}; column temperature—25 °C; back pressure—130 bar; injection volume—2.0 μL; Table S1: Effect of SFC conditions on the parameters of chromatographic separation of analytes (k—retention factor; $W_{1/2}$—peak width at half-height; α—selectivity; R—resolution); Table S2: The intra-day and inter-day precision of the developed approach; Table S3: In silico predicted toxicity and mutagenicity of analytes.

Author Contributions: Conceptualization, N.V.U. and D.S.K.; methodology, N.V.U. and D.V.O.; formal analysis, N.V.U. and D.V.O.; investigation, D.V.O., S.A.V. and D.I.F.; resources, N.V.U. and D.S.K.; writing—original draft preparation, D.V.O.; writing—review and editing, N.V.U. and D.S.K.; supervision, D.S.K.; project administration, N.V.U. and D.S.K.; funding acquisition, N.V.U. and D.S.K. All authors have read and agreed to the published version of the manuscript.

Funding: This research was funded by Russian Foundation for Basic Research (grants No. 20-03-00416-A and 19-33-60011) and the Ministry of Science and Higher Education of the Russian Federation (project No. 0793-2020-0007) as part of the study of peat bog soil pollution.

Institutional Review Board Statement: Not applicable.

Informed Consent Statement: Not applicable.

Data Availability Statement: Not applicable.

Acknowledgments: An instrumentation of the Core Facility Center "Arktika" of the Northern (Arctic) Federal University was used in this study.

Conflicts of Interest: The authors declare no conflict of interest.

Sample Availability: Samples are not available from the authors.

References

1. Buryak, A.K.; Serdyuk, T.M. Chromatography—mass spectrometry in aerospace industry. *Russ. Chem. Rev.* **2013**, *82*, 369–392. [CrossRef]
2. Hu, C.; Zhang, Y.; Zhou, Y.; Liu, Z.-F.; Feng, X.-S. Unsymmetrical dimethylhydrazine and related compounds in the environment: Recent updates on pretreatment, analysis, and removal techniques. *J. Hazard. Mater.* **2022**, *432*, 128708. [CrossRef] [PubMed]
3. Ul'yanovskii, N.V.; Kosyakov, D.S.; Pikovskoi, I.I.; Khabarov, Y.G. Characterisation of oxidation products of 1,1-dimethylhydrazine by high-resolution orbitrap mass spectrometry. *Chemosphere* **2017**, *174*, 66–75. [CrossRef] [PubMed]
4. Kosyakov, D.S.; Ul'yanovskii, N.V.; Pikovskoi, I.I.; Kenessov, B.; Bakaikina, N.V.; Zhubatov Zh Lebedev, A.T. Effects of oxidant and catalyst on the transformation products of rocket fuel 1,1-dimethylhydrazine in water and soil. *Chemosphere* **2019**, *228*, 335–344. [CrossRef]
5. Kenessov, B.; Alimzhanova, M.; Sailaukhanuly, Y.; Baimatova, N.; Abilev, M.; Batyrbekova, S.; Carlsen, L.; Tulegenov, A.; Nauryzbayev, M. Transformation products of 1,1-dimethylhydrazine and their distribution in soils of fall places of rocket carriers in Central Kazakhstan. *Sci. Total Environ.* **2012**, *427–428*, 78–85. [CrossRef]
6. Rodin, I.A.; Moskvin, D.N.; Smolenkov, A.D.; Shpigun, O.A. Transformations of asymmetric dimethylhydrazine in soils. *Russ. J. Phys. Chem. A* **2008**, *82*, 911–915. [CrossRef]
7. Milyushkin, A.L.; Birin, K.P.; Matyushin, D.D.; Semeikin, A.V.; Iartsev, S.D.; Karnaeva, A.E.; Uleanov, A.V.; Buryak, A.K. Isomeric derivatives of triazoles as new toxic decomposition products of 1,1-dimethylhydrazine. *Chemosphere* **2019**, *217*, 95–99. [CrossRef]
8. Kenessov, B.N.; Koziel, J.A.; Grotenhuis, T.; Carlsen, L. Screening of transformation products in soils contaminated with unsymmetrical dimethylhydrazine using headspace SPME and GC-MS. *Anal. Chim. Acta* **2010**, *674*, 32–39. [CrossRef]
9. Smolenkov, A.D.; Rodin, I.A.; Shpak, A.V.; Shpigun, O.A. 1-Formyl-2,2-dimethylhydrazine as a new decomposition product of 1,1-dimethylhydrazine. *Int. J. Environ. Anal. Chem.* **2007**, *87*, 351–359. [CrossRef]
10. Yegemova, S.; Bakaikina, N.V.; Kenessov, B.; Koziel, J.A.; Nauryzbayev, M. Determination of 1-methyl-1H-1,2,4-triazole in soils contaminated by rocket fuel using solid-phase microextraction, isotope dilution and gas chromatography-mass spectrometry. *Talanta* **2015**, *143*, 226–233. [CrossRef]
11. Kosyakov, D.S.; Ul'yanovskii, N.V.; Bogolitsyn, K.G.; Shpigun, O.A. Simultaneous determination of 1,1-dimethylhydrazine and products of its oxidative transformations by liquid chromatography–tandem mass spectrometry. *Int. J. Environ. Anal. Chem.* **2014**, *94*, 1254–1263. [CrossRef]
12. Rodin, I.A.; Anan'eva, I.A.; Smolenkov, A.D.; Shpigun, O.A. Determination of the products of the oxidative transformation of unsymmetrical dimethylhydrazine in soils by liquid chromatography/mass spectrometry. *J. Anal. Chem.* **2010**, *65*, 1405–1410. [CrossRef]
13. Huang, D.; Liu, X.; Wang, X.; Huang, Z.; Xie, Z.; Wang, H. Investigation on the compositions of unsymmetrical dimethylhydrazine treatment with different oxidants using solid-phase micro-extraction-gas chromatography-mass spectrometer. *R. Soc. Open Sci.* **2019**, *6*, 190263. [CrossRef] [PubMed]
14. Smirnov, R.S.; Rodin, I.A.; Smolenkov, A.D.; Shpigun, O.A. Determination of the products of the transformation of unsymmetrical dimethylhydrazine in soils using chromatography/mass spectrometry. *J. Anal. Chem.* **2010**, *65*, 1266–1272. [CrossRef]
15. Ul'yanovskii, N.V.; Lakhmanov, D.E.; Pikovskoi, I.I.; Falev, D.I.; Popov, M.S.; Kozhevnikov, A.Y.; Kosyakov, D.S. Migration and transformation of 1,1-dimethylhydrazine in peat bog soil of rocket stage fall site in Russian North. *Sci. Total Environ.* **2020**, *72615*, 138483. [CrossRef]

16. ACD/Labs Percepta software, Advanced Chemistry Development Inc.: Toronto, ON, Canada. Available online: http://www.acdlabs.com/home/ (accessed on 23 July 2022).
17. Smolenkov, A.D.; Shpigun, O.A. Direct liquid chromatographic determination of hydrazines: A review. *Talanta* **2012**, *102*, 93–100. [CrossRef]
18. Agrawal, R.; Belemkar, S.; Bonde, C. Orthogonal Separations in Reversed-Phase Chromatography. *Chromatographia* **2018**, *81*, 565–573. [CrossRef]
19. Lee, J.W.; Nagai, T.; Gotoh, N.; Fukusaki, E.; Bamba, T. Profiling of regioisomeric triacylglycerols in edible oils by supercritical fluid chromatography/tandem mass spectrometry. *J. Chromatogr. B* **2014**, *966*, 193–199. [CrossRef]
20. Regalado, E.L.; Schafer, W.; McClain, R.; Welch, C.J. Chromatographic resolution of closely related species: Separation of warfarin and hydroxylated isomers. *J. Chromatogr. A* **2013**, *1314*, 266–275. [CrossRef]
21. Xhaferaj, M.; Naegele, E.; Parr, M.K. Ion exchange in supercritical fluid chromatography tandem mass spectrometry (SFC-MS/MS): Application for polar and ionic drugs and metabolites in forensic and anti-doping analysis. *J. Chromatogr. A* **2020**, *1614*, 460726. [CrossRef]
22. Wolrab, D.; Frühauf, P.; Gerner, C. Direct coupling of supercritical fluid chromatography with tandem mass spectrometry for the analysis of amino acids and related compounds: Comparing electrospray ionization and atmospheric pressure chemical ionization. *Anal. Chim. Acta* **2017**, *981*, 106–115. [CrossRef] [PubMed]
23. Antonelli, M.; Holčapek, M.; Wolrab, D. Ultrahigh-performance supercritical fluid chromatography—mass spectrometry for the qualitative analysis of metabolites covering a large polarity range. *J. Chromatogr. A* **2022**, *1665*, 462832. [CrossRef] [PubMed]
24. Grand-Guillaume Perrenoud, A.; Veuthey, J.-L.; Guillarme, D. Coupling state-of-the-art supercritical fluid chromatography and mass spectrometry: FROM hyphenation interface optimization to high-sensitivity analysis of pharmaceutical compounds. *J. Chromatogr. A* **2014**, *1339*, 174–184. [CrossRef] [PubMed]
25. Akbal, L.; Hopfgartner, G. Hyphenation of packed column supercritical fluid chromatography with mass spectrometry: Where are we and what are the remaining challenges? *Anal. Bioanal. Chem.* **2020**, *412*, 6667–6677. [CrossRef]
26. Van de Velde, B.; Guillarme, D.; Kohler, I. Supercritical fluid chromatography–Mass spectrometry in metabolomics: Past, present, and future perspectives. *J. Chromatogr. B* **2020**, *1161*, 122444. [CrossRef]
27. Ovchinnikov, D.V.; Ul'yanovskii, N.V.; Falev, D.I.; Kosyakov, D.S. Supercritical Fluid Chromatography–Mass-Spectrometry of Nitrogen-Containing Compounds: Atmospheric Pressure Ionization. *J. Anal. Chem.* **2021**, *76*, 1624–1634. [CrossRef]
28. Gohres, J.L.; Marin, A.T.; Jie, L.; Liotta, C.L.; Eckert, C.A. Spectroscopic investigation of alkylcarbonic acid formation and dissociation in CO 2-Expanded alcohols. *Ind. Eng. Chem. Res.* **2009**, *48*, 1302–1306. [CrossRef]
29. Lin, H.-W.; Yen, C.H.; Hsu, H.; Tan, C.-S. CO2 promoted hydrogenolysis of benzylic compounds in methanol and water. *RSC Adv.* **2013**, *3*, 17222–17227. [CrossRef]
30. Falev, D.I.; Ovchinnikov, D.V.; Voronov, I.S.; Faleva, A.V.; Ul'yanovskii, N.V.; Kosyakov, D.S. Supercritical Fluid Chromatography—Tandem Mass Spectrometry for Rapid Quantification of Pentacyclic Triterpenoids in Plant Extracts. *Pharmaceuticals* **2022**, *15*, 629. [CrossRef]
31. West, C.; Lemasson, E.; Nagai, K.; Nagai, K.; Shibata, T.; Franco, P.; Hennig, P.; Lesellier, E. Characterization of Novel Polymer-Based Pyridine Stationary Phases for Supercritical Fluid Chromatography. *Chromatographia* **2019**, *82*, 143–152. [CrossRef]
32. Cazenave-Gassiot, A.; Boughtflower, R.; Caldwell, J.; Hitzel, L.; Holyoak, C.; Lane, S.; Oakley, P.; Pullen, F.; Richardson, S.; Langley, G.J. Effect of increasing concentration of ammonium acetate as an additive in supercritical fluid chromatography using CO2-methanol mobile phase. *J. Chromatogr. A* **2009**, *1216*, 6441–6450. [CrossRef] [PubMed]
33. Ovchinnikov, D.V.; Pokrovskiy, O.I.; Kosyakov, D.S.; Bogolitsyn, K.G.; Ul'yanovskii, N.V.; Falev, D.I. Evaluation of temperature and pressure effects on retention in supercritical fluid chromatography on polar stationary phases. *J. Chromatogr. A* **2020**, *1610*, 460600. [CrossRef] [PubMed]
34. Kosyakov, D.S.; Ul'yanovskii, N.V.; Pokryshkin, S.A.; Lakhmanov, D.E.; Shpigun, O.A. Rapid determination of 1,1-dimethylhydrazine transformation products in soil by accelerated solvent extraction coupled with gas chromatography–tandem mass spectrometry. *Int. J. Environ. Anal. Chem.* **2015**, *95*, 1321–1337. [CrossRef]
35. Ul'yanovskii, N.V.; Kosyakov, D.S.; Popov, M.S.; Shavrina, I.S.; Ivakhnov, A.D.; Kenessov, B.; Lebedev, A.T. Rapid quantification and screening of nitrogen-containing rocket fuel transformation products by vortex assisted liquid-liquid microextraction and gas chromatography—high-resolution Orbitrap mass spectrometry. *Microchem. J.* **2021**, *171*, 106821. [CrossRef]
36. U.S. EPA MRID 41514401. Determination of 1,3-Dimethylurea in Soil by Gas Chromatography. US Environmental Protection Agency. 1989. Available online: https://www.epa.gov/sites/default/files/2015-01/documents/415144-01-s.pdf (accessed on 23 July 2022).
37. Beltrami, R.T.; Bissell, E.R. Some Methylhydrazonium Salts; An Improved Synthesis of Tetramethylhydrazine. *J. Am. Chem. Soc.* **1956**, *78*, 2467–2468. [CrossRef]
38. *Technological regulations. Detoxification of Soils Contaminated with Asymmetric Dimethylhydrazine and Products of its Chemical Transformation by a Combined Method*; National Space Agency of the Republic of Kazakhstan: Almaty, Kazakhstan, 2012; p. 16. (In Russian)

MDPI AG
Grosspeteranlage 5
4052 Basel
Switzerland
Tel.: +41 61 683 77 34

Molecules Editorial Office
E-mail: molecules@mdpi.com
www.mdpi.com/journal/molecules

www.ingramcontent.com/pod-product-compliance
Lightning Source LLC
Chambersburg PA
CBHW062052220926
51850CB00010B/2742
* 9 7 8 3 7 2 5 8 6 7 8 6 8 *